Algebra I
Part A

Mathematics

Nancy L. McGraw

Bright Ideas Press, LLC
Cleveland, Ohio

Simple Solutions Algebra I Part A

Printed in the United States of America

ISBN-13: 978-1-934210-18-5
ISBN-10: 1-934210-18-8

Cover Design: Dan Mazzola
Editor: Kimberly A. Dambrogio

Welcome to Simple Solutions

Note to the Student:

This workbook will give you the opportunity to practice skills you have learned in previous grades. By practicing these skills each day, you will gain confidence in your math ability.

Using this workbook will help you understand math concepts easier and for many of you, it will give you a more positive attitude toward math in general.

In order for this program to help you be successful, it is extremely important that you do a lesson every day. It is also important that you check your answers and ask your teacher for help with the problems you didn't understand or that you did incorrectly.

If you put forth the effort, Simple Solutions will change your opinion about math forever.

Lesson #1

1. $-19 + (-26) = ?$
2. $|-29| = ?$
3. $\sqrt{2{,}025} = ?$
4. $63 + (-46) = ?$
5. $3\frac{1}{3} + 4\frac{1}{5} = ?$
6. Solve for a. $a - 12 = 56$
7. You are building a fence 13 feet long. On Monday, you build $4\frac{1}{3}$ feet of the fence, and, on Tuesday, $5\frac{3}{4}$ feet. How much of the fence do you still need to build?

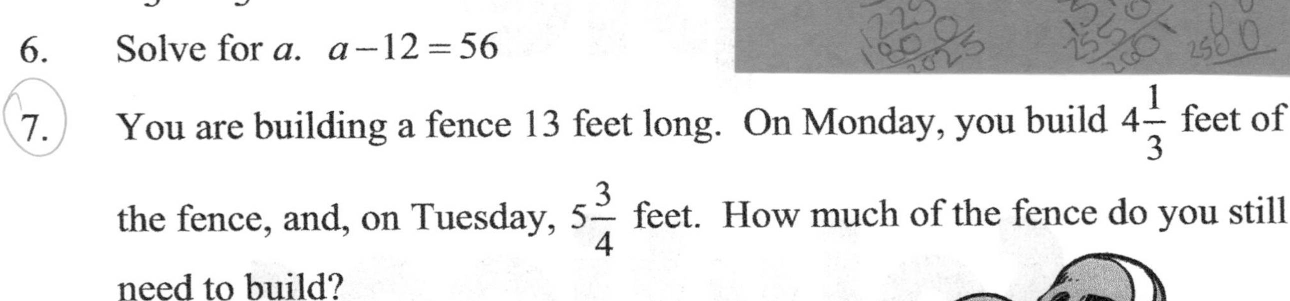

8. $0.06 \times 0.05 = ?$
9. $\frac{81}{-9} = ?$
10. Simplify. $6a + 4b + 8 + 3b - 4a + 9$
11. $-22 - (-16) = ?$
12. What is the value of x? $\frac{x}{5} = 15$
13. $-6(-5) = ?$
14. Write the ratio *six to seven* in two other ways.
15. Find the circumference of the circle.

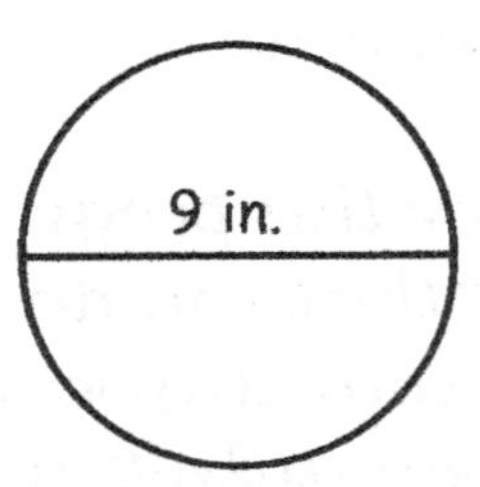

16. Evaluate $\frac{3ab}{2}$ when $a = 2$ and $b = 3$.
17. Change $\frac{4}{5}$ to a decimal and to a percent.
18. $30 \div 5 \cdot 6 + 14 - 10 \div 2 = ?$
19. Find the area of a square if a side measures 12 cm.
20. Write an algebraic expression for *the difference of a number and 10.*

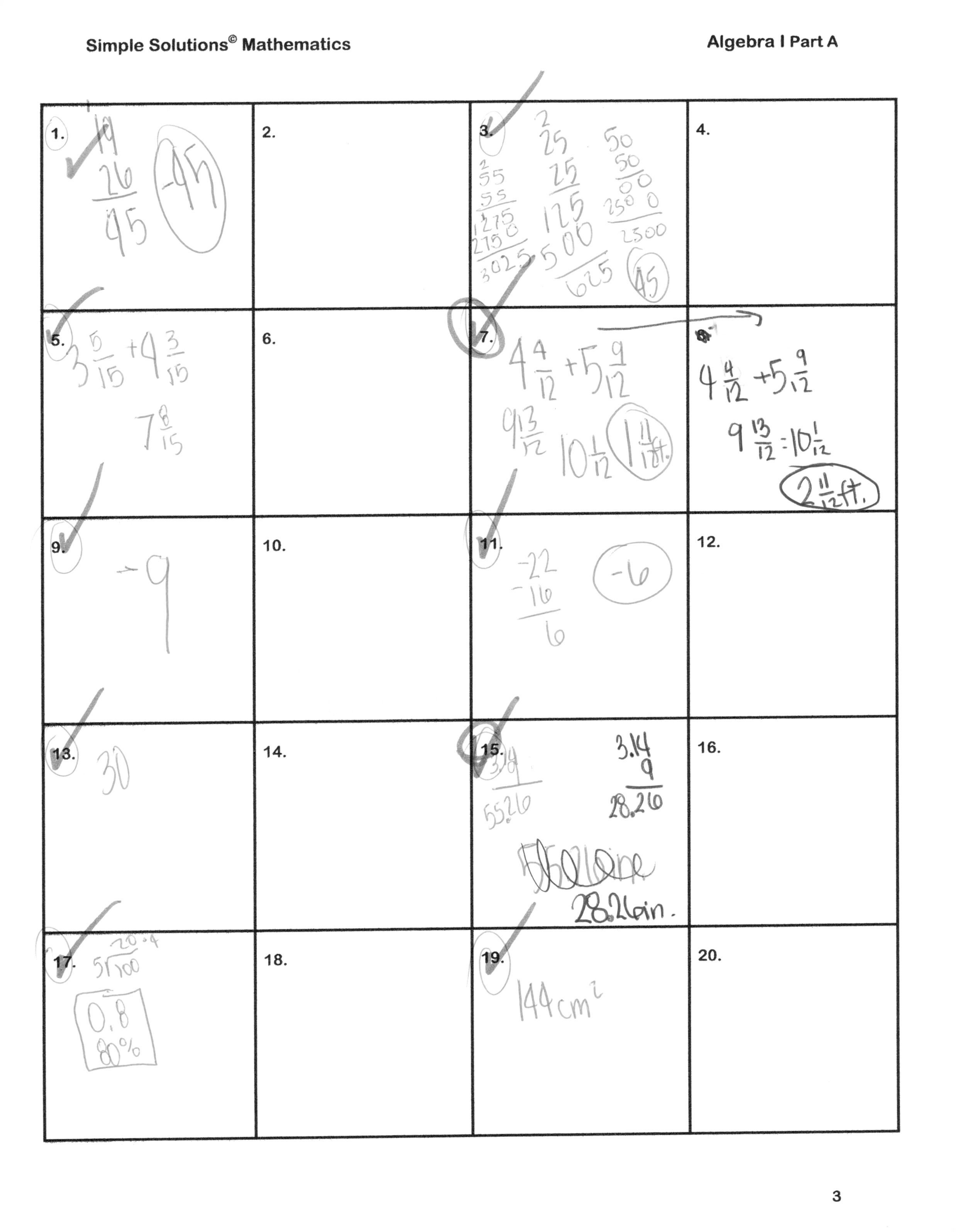

1.	2.	3.	4.
5.	6.	7.	8.
9.	10.	11.	12.
13.	14.	15.	16.
17.	18.	19.	20.

Lesson #2

1. How many sides does a pentagon have?
2. $96 \times 57 = ?$
3. $50 \div 5 \cdot 9 - 45 + 5 = ?$
4. Solve for b. $b + 42 = -86$
5. $436{,}816 + 598{,}444 = ?$
6. What is the value of $3xy + 2x$ when $x = 2$ and $y = 3$?
7. Find the value of a. $2a = 36$
8. Make a factor tree for 45.
9. Round 56,477,812 to the nearest hundred thousand.
10. $-55 - (-39) = ?$
11. Combine like terms. $6x + 12 + 4y + 7x + 15$
12. $36.24 + 19.8 = ?$
13. What is the value of x? $\frac{x}{12} = 9$
14. $8 - 3\frac{4}{5} = ?$
15. $-6(-2)(3) = ?$
16. Find the LCM of $6x^2yz^4$ and $15xy^3z^2$.
17. $64 + (-26) = ?$
18. After taxes, Lamar brings home $500 per week in his pay. He would like to save enough money to put a $2,000 down payment on a new motorcycle. If Lamar saves 50% of his weekly salary toward the down payment on a new motorcycle, how many weeks will it take him to reach his goal?
19. Write an algebraic phrase to represent *a number divided by 8*.
20. What value of x makes the two fractions equivalent? $\frac{9}{5} = \frac{108}{x}$

1.	2. 96 × 51 = 96 + 4800 = 5412 → **5412**	3. $b + 42 = -86$; -42 -42; $-86 + 42$ → **$b = 128$**	4.
5.	6. $3 \cdot 2 \cdot 3 + 2 \cdot 2$; $18 + 4$ → **22**	7.	8. 45 = 9 · 5 = 3 · 3 · 5
9.	10. 55 − 39 = 16 → **−16**	11.	12. 36.24 + 19.8 = **56.04**
13.	14. $\frac{40}{5} - \frac{19}{5}$ $\frac{21}{5}$ → **$4\frac{1}{5}$**	15. 12 · 3 → **36**	16. $6x^2yz^4$; $15xy^3z^2$; $30x^2yz^2$
17.	18. **8**	19.	20. $\frac{9}{5} = \frac{108}{x}$; $9x = 540$; $60 = x$

Lesson #3

1. $93 - (-18) = ?$
2. How long are the sides of a square if its area is 100 ft^2?
3. $\sqrt{2,601} = ?$
4. Write 35% as a decimal and as a reduced fraction.
5. Solve for a. $\frac{a}{4} = 16$

6. $|54| = ?$
7. $-33 + (-47) = ?$
8. Find the value of a. $5a = 75$
9. It is 4:20 now. What time will it be in 6 hours and 15 minutes?
10. $\frac{9}{12} \bigcirc \frac{7}{11}$
11. $3,080 \div 4 = ?$

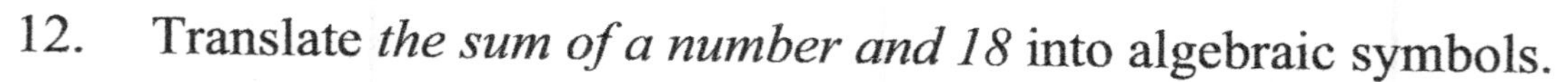

12. Translate *the sum of a number and 18* into algebraic symbols.
13. $\frac{-54}{9} = ?$
14. Mary is 3 times as old as her brother, Josh. If Josh is 6 years old, how old will Mary be on her next birthday?
15. Round 46,877,641 to the nearest thousand.
16. $\frac{3}{5} \times \frac{15}{18} = ?$
17. Put these integers in order from least to greatest. $-16, 0, 2, -24, 16$

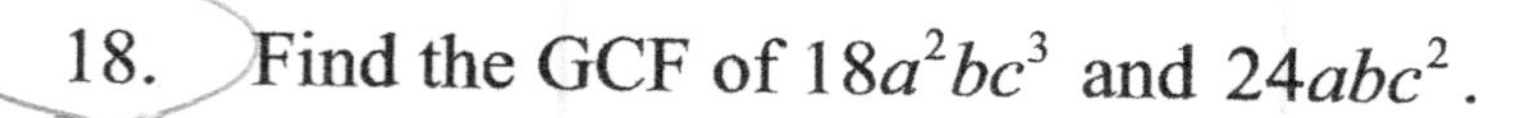

18. Find the GCF of $18a^2bc^3$ and $24abc^2$.
19. Find the median of 17, 24, 98, 67 and 42.
20. Give the coordinates for points F, A, and D.

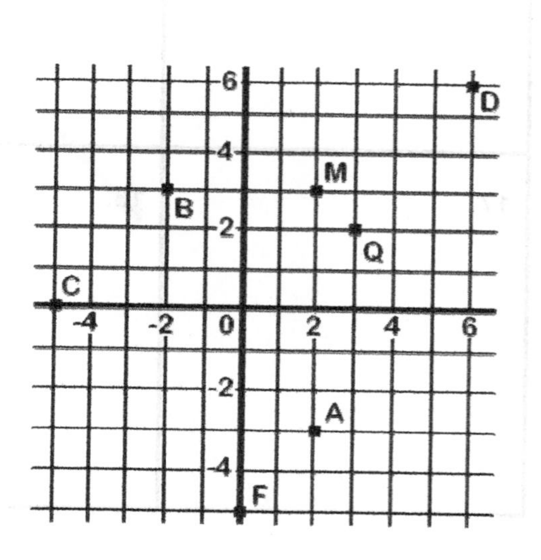

1.	2.	3. 50 × 50 = 2500; 52 × 52 = 2704; 51 × 51 = 2601; (51)	4.
5.	6. 54	7.	8. $\frac{5a}{5} = \frac{75}{5}$ $a = 15$
9. 4:20 + 6 15 = 10:35; 10:35	10.	11.	12.
13.	14. $3 \cdot 6 = 18 + 1 =$ (19 yrs.)	15. 46,878,000	16. $\frac{1}{5}$, $\frac{153}{186} = \frac{3}{6} =$ ($\frac{1}{2}$)
17. −24, −16, [illegible], 16	18. $6abc^2$	19.	20. F(0, −5) a(2, −3) D(6, 6)

Lesson #4

1. On his way from Toledo to Cincinnati, Mr. Smith drove at an average rate of 65 miles per hour. If his trip took 5 hours, how far did he travel?

2. Find the area of a circle with a radius of 4 inches. (Hint: $A = \pi r^2$)

3. $-8(-4)(2) = ?$

4. A triangle with 2 congruent sides is a(n) _______ triangle.

5. $-88 - (-49) = ?$

6. $0.6 - 0.2475 = ?$

7. $2\frac{7}{10} - 1\frac{3}{5} = ?$

8. Solve for a. $\frac{a}{7} = 21$

9. Solve the proportion for x. $\frac{x}{5} = \frac{24}{60}$

10. How many quarts are in 6 gallons?

11. $-16 + (-14) + 11 = ?$

12. $36 \div 6 \cdot 2 + 10 - 5 = ?$

13. List the first 5 prime numbers.

14. Find the value of n. $n - 15 = 35$

15. The ratio of lions to giraffes living in the grasslands was 13 to 2. If there were 52 lions, how many giraffes lived in the grasslands?

16. Write $\frac{13}{20}$ as a decimal and as a percent.

17. Solve for x. $4x = 64$

18. $365{,}812 + 132{,}999 = ?$

19. What number is 60% of 90?

20. Simplify by combining like terms. $5a + 7 + 3b + 8a + 10 + 14a$

1.	2.	3.	4.
5.	6.	7.	8.
9.	10.	11.	12.
13.	14.	15.	16.
17.	18.	19.	20.

Lesson #5

1. $-55-(-18)=?$

2. Solve this equation to find the value of x. $3x-7=14$

3. $7\frac{1}{2}+3\frac{2}{5}=?$

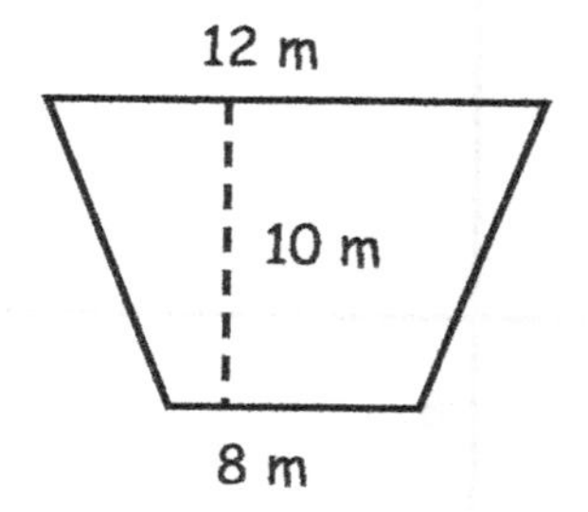

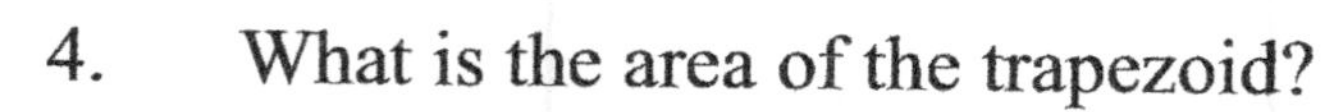

4. What is the area of the trapezoid?

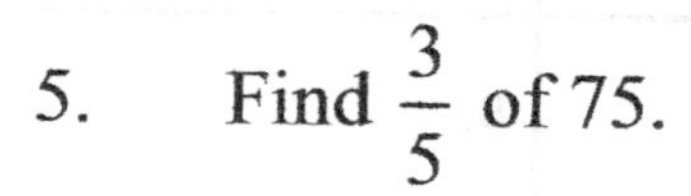

5. Find $\frac{3}{5}$ of 75.

6. Solve for x. $\frac{x}{4}=15$

7. Give the name of each shape. a) b)

8. $0.003 \times 0.07=?$

9. Write 65% as a decimal and as a reduced fraction.

10. How many hamburgers can Melinda make from 5 pounds of hamburger if she uses $\frac{1}{4}$ pound of meat per hamburger?

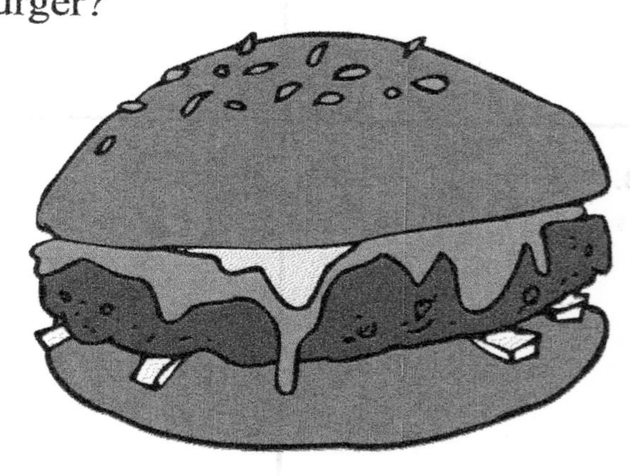

11. $36 \div 6 \cdot 4-4+10 \div 5=?$

12. $\sqrt{9{,}801}=?$

13. 25% of 80 is what number?

14. Find the GCF of $16x^2yz^3$ and $15xy^3z^2$.

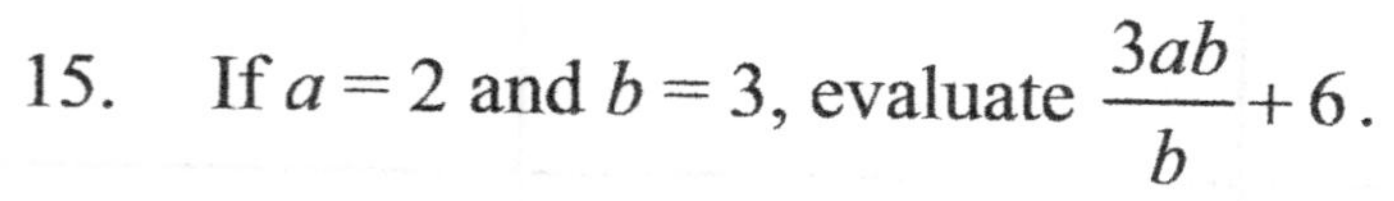

15. If $a=2$ and $b=3$, evaluate $\frac{3ab}{b}+6$.

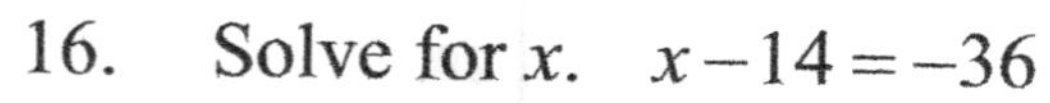

16. Solve for x. $x-14=-36$

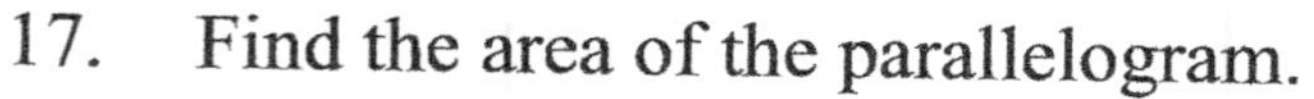

17. Find the area of the parallelogram.

6 m

7 m

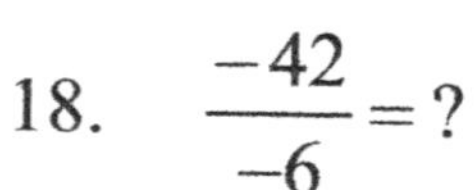

18. $\frac{-42}{-6}=?$

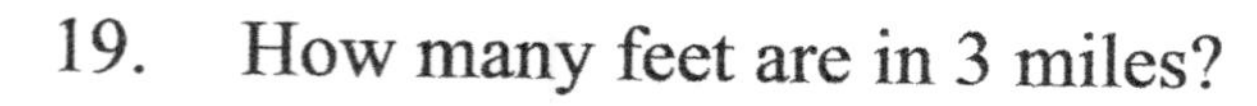

19. How many feet are in 3 miles?

20. Solve for x. $7x=105$

1.	2.	3.	4.
5.	6.	7.	8.
9.	10.	11.	12.
13.	14.	15.	16.
17.	18.	19.	20.

Lesson #6

1. Write $6 \times 6 \times 6 \times 6 \times 6 \times 6$ using a base and an exponent.

2. $41 + (-23) = ?$

3. $\frac{-21}{3} = ?$

4. $\sqrt{784} = ?$

5. Solve for x. $\frac{5}{8} = \frac{x}{120}$

6. Solve for x. $3x = 48$

7. When $x = 3$ and $y = 4$, what is the value of $4xy + x$?

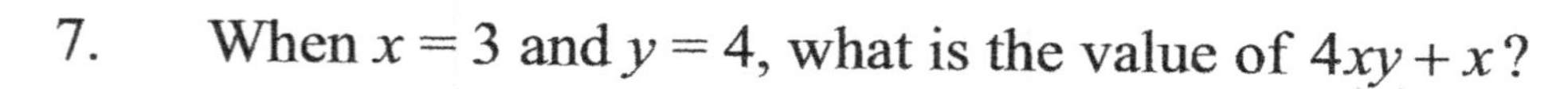

8. Find the volume of the rectangular prism.

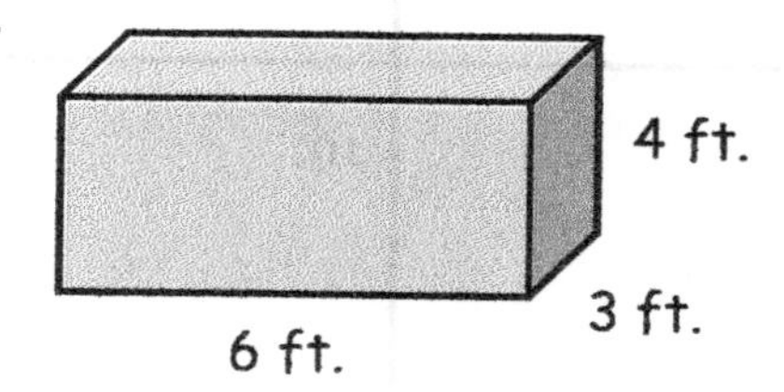

9. Simplify. $\frac{8a^2bc^3}{12ab^2c^5}$

10. $|-91| = ?$

11. What value of a makes the equation true? $3a + 12 = 6a - 24$

12. Solve for b. $b + 16 = -52$

13. Simplify by combining like terms. $4a + 2b + 8 + 5a + 6b - 6$

14. $11 - 7\frac{3}{7} = ?$

15. Put these integers in decreasing order. $-56,\ 32,\ -3,\ 46$

16. $10 \cdot 5 \div 2 + 5 - 3 = ?$

17. Write $\frac{12}{18}$ in simplest form.

18. Mr. Giles has 6 horses; each horse eats about 85 pounds of feed per year. How many pounds of feed will 6 horses eat in $1\frac{1}{2}$ years?

19. $63 - (-19) = ?$

20. Solve for a. $a - 17 = -52$

1. 6^6	2. $41 - 23 = 18$; 18	3. $3\overline{)21}$ = 7; −7	4. 30 × 30 = 900; 26 × 26 = 676; 25 × 25 = 625; 28
5. $\frac{5}{8} = \frac{x}{120}$; $8x = 600$; $x = 65$	6. $\frac{3x}{3} = \frac{48}{3}$; $x = 16$	7. $4 \cdot 3 \cdot 4 + 3$; $48 + 3$; 51	8. $6 \cdot 3 \cdot 4$; $6 \cdot 12$; 72
9. $\frac{2abc^3}{3abc^3}$	10. 91	11. $3a + 12 = 6a - 24$; $24 + 12 = 3a$; $\frac{36}{3} = \frac{3a}{3}$; $a = 12$	12. $52 + 16$; −68
13. $4a + 2b + 8 + 5a + 6b - 6$; $9a + 8b + 2$	14. $\frac{77}{7} - \frac{52}{7}$; $\frac{25}{7} = 3\frac{4}{7}$	15. 46, 32, −3, −56	16. $40 \cdot 5 \div 2 + 5 - 3$; $50 \div 2 + 5 - 3$; $25 + 5 - 3$; $30 - 3$; 27
17. $\frac{6}{9} = \frac{2}{3}$	18. 42.5 × 6 = 255; 85 × 6 = 510; 255 + 510 = 765 lbs.	19. 63 + 19; 82	20. $a - 17 = -52 + 17$; $a = -35$

Lesson #7

1. Find the value of $3a+2b-3$ when $a=5$ and $b=4$.

2. Write $\frac{4}{25}$ as a decimal and as a percent.

3. Solve for b. $b+16=-42$

4. $16 \div 4 \cdot 5 + 10 \div 2 - 5 = ?$

5. $0.13 \times 0.09 = ?$

6. Solve for x. $\frac{3}{5}x=60$ (Hint: Multiply by the reciprocal of $\frac{3}{5}$.)

7. Jake is going to bowl three games. If he scores 125 and 113 points in the first two games, how many points will Jake need in the third game so that his average score is 126 points?

8. $-65-(-39)=?$

9. $-4(4)=?$

10. Solve the proportion for x. $\frac{x}{8}=\frac{48}{96}$

11. Solve for h. $h+52=-80$

12. Simplify. $5(3x-2y+7)$

13. $16\frac{1}{2}-10\frac{3}{4}=?$

14. $\sqrt{0.0004}=?$

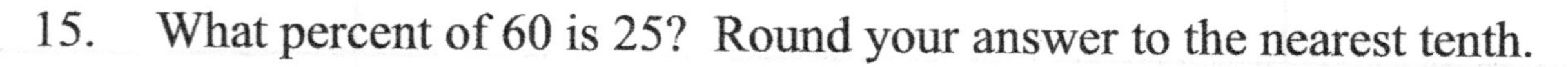

15. What percent of 60 is 25? Round your answer to the nearest tenth.

16. $72+(-36)=?$

17. Solve for a. $a-17=-13$

18. $62{,}378+57{,}984=?$

19. Find the value of x. $4x+9=7x$

20. Round 36,776,210 to the nearest million.

1. $3\cdot5+2\cdot4-3$ $15+8-3$ $23-3$ 20	2. $4\cdot4=16\%$ 16% $25\overline{)100}$ = 4	3. $b+16=-42$ $B=-42+-16$ $B=58$	4. $16\div4\cdot5+10\div2-5$ $4\cdot5+10\div2-5$ $20+10\div2-5$ $20+5-5$ $25-5$ 20
5. 13 09 1.17 1.17	6. $\frac{3}{5}x=60$ $x=60\cdot\frac{5}{3}$ $x=100$	7. $(125+113)\div2$ $238\div2$ 119 $x\div3=126\cdot3$ 140	8. 65 39 26 −26
9. −16	10. $\frac{x}{8}=\frac{48}{96}$ $\frac{384}{96}=\frac{96x}{96}$ $x=4$	11. $H+52=-80+52$ $H=-132$	12. $15x-10y+35$
13. $15\frac{6}{4}-10\frac{3}{4}$ $5\frac{3}{4}$	14. .02 .02 .02 .0004	15. $\frac{25}{100}=\frac{x}{100}$ $2500=100x$ $41\frac{2}{3}\%$	16. 72 −36 36 36
17. $9-17=-13+17$ $+17$ $9=4$	18. 62378 57984 120352 120352	19. $4x+9=7x-4x$ $\frac{9}{3}=\frac{3x}{3}$ $3=x$	20. 37,000,000

Lesson #8

1. $37{,}897 - 19{,}543 = ?$

2. Solve for a. $a - 28 = 90$

3. Calculate the area of a triangle if its base is 15 cm long and its height is 6 cm.

4. $\dfrac{-44}{11} = ?$

5. Simplify. $7(3x - 5y + 9) + 3(4x - 7)$

6. $-66 - (-49) = ?$

7. $9\frac{2}{5} + 3\frac{1}{8} = ?$

8. Write 0.65 as a reduced fraction.

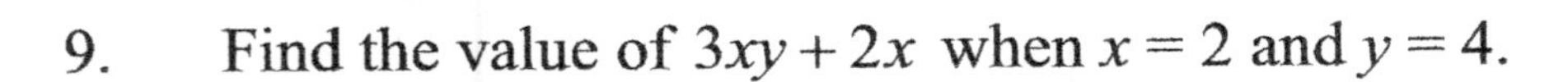

9. Find the value of $3xy + 2x$ when $x = 2$ and $y = 4$.

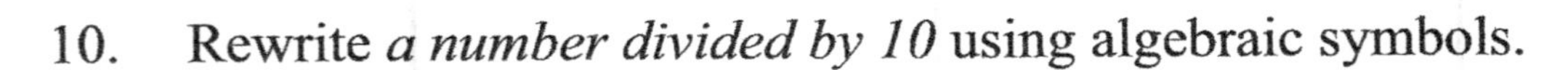

10. Rewrite *a number divided by 10* using algebraic symbols.

11. $16 \cdot 2 \div 4 + 12 - 14 \div 2 = ?$

12. $-7(-4)(2) = ?$

13. $13 - 8\frac{2}{7} = ?$

14. $-|36| = ?$

15. $92 + (-37) = ?$

16. Marissa went to the store with \$20 and returned home with \$5.36. If she bought only 3 gallons of ice cream, what was the cost of each gallon?

17. $0.7 - 0.1346 = ?$

18. Solve for x. $\dfrac{x}{5} = 12$

19. What is the value of x? $9x - 9 = 12x$

20. Round 37.4625 to the nearest tenth.

1.	2.	3.	4.
5.	6.	7.	8.
9.	10.	11.	12.
13.	14.	15.	16.
17.	18.	19.	20.

Lesson #9

1. $\begin{pmatrix} 7 & 8 & -6 \\ 0 & 2 & -2 \end{pmatrix} + \begin{pmatrix} -3 & 0 & -4 \\ 9 & 7 & 1 \end{pmatrix} = ?$
2. Solve to find the value of a. $a - 22 = -61$
3. $75 \times 24 = ?$
4. What value of x makes these fractions equivalent? $\frac{3}{7} = \frac{x}{105}$
5. $-|-54| = ?$
6. Find the LCM of $16x^2y^5z^4$ and $21xy^3z^2$.
7. Evaluate $7xy - 2y + 4$ when $x = 2$ and $y = 3$.
8. Find the area of the parallelogram below.

9. Solve for b. $b + 12 = -57$
10. Put $\frac{18}{24}$ in simplest form.
11. $-55 + (-21) = ?$
12. $3 \cdot 3 \cdot 4 \div 9 + 14 - 2 \div 1 = ?$
13. $-7(5) = ?$
14. Solve for x. $\frac{x}{6} = 14$
15. $46 - (-24) = ?$
16. $80{,}000 - 26{,}474 = ?$
17. The enrollment of a middle school is 360 students. If the office reports that $\frac{3}{10}$ of the students are in sixth grade and $\frac{7}{20}$ are in seventh grade, how many students are in eighth grade?
18. $\frac{12}{15} \times \frac{5}{8} = ?$
19. Simplify. $3(2a + 4b + 7) + 6a$
20. Solve for x. $\frac{2}{5}x = 10$

1.	2.	3.	4.
5.	6.	7.	8.
9.	10.	11.	12.
13.	14.	15.	16.
17.	18.	19.	20.

Lesson #10

1. If $x = 3$, $y = 6$ and $z = 5$, find the value of $x(y \cdot z)$.

2. Evaluate. $(-3)^4$

3. Simplify. $5(3x - 4y - 8) + 4x - 7$

4. Solve for a. $5a + 4 = 14$

5. Find the area of a triangle whose base is 14 inches and height is 6 inches.

6. Translate *a number decreased by 13* into an algebraic phrase.

7. $77 - (-54) = ?$

8. $60{,}000 - 29{,}312 = ?$

9. Solve for a. $a - 31 = -73$

10. The ratio of polar bears to caribou at the game preserve was 5 to 3. If there were 125 polar bears, how many caribou were there?

11. 70% of 90 is what number?

12. $92 + (-53) = ?$

13. Solve for x. $\frac{x}{5} = 13$

14. $-8(-7) = ?$

15. Find the value of b. $b + 13 = 39$

16. $16\frac{2}{3} + 24\frac{3}{5} = ?$

17. Three-fifths of the students passed their test. What percent of the students did not pass?

18. Solve for x. $3x = 69$

19. $4 \cdot 5 \cdot 3 \div 6 + 25 - 2^2 = ?$

20. $5.4 \div 4.5 = ?$

1.	2.	3.	4.
5.	6.	7.	8.
9.	10.	11.	12.
13.	14.	15.	16.
17.	18.	19.	20.

Lesson #11

1. Solve for x. $x + 36 = 122$
2. Write $\frac{1}{5}$ as a decimal and as a percent.
3. $86 - (-39) = ?$
4. $\frac{-224}{8} = ?$
5. How many feet are in 5 miles?
6. $8^2 - \sqrt{81} = ?$
7. $6\frac{3}{7} - 2\frac{6}{7} = ?$
8. Solve for a. $\frac{a}{5} = 14$
9. Simplify. $5(2a + 6b - 6) - 4(2a - 5)$
10. What is the value of x? $2x - 8 = 16$
11. $42.3 - 5.787 = ?$
12. Find the value of $\frac{3xy}{y} + 2x$ when $x = 2$ and $y = 5$.

12 cm
6 cm
10 cm

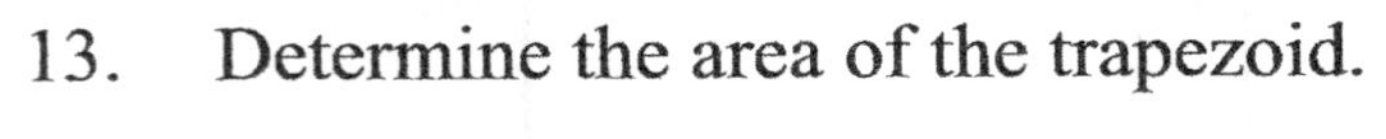

13. Determine the area of the trapezoid.

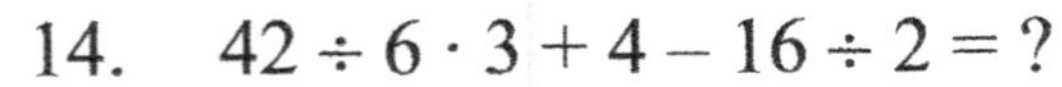

14. $42 \div 6 \cdot 3 + 4 - 16 \div 2 = ?$
15. Calculate the area of a circle if its radius is 5 feet.
16. Express this phrase in algebraic symbols: *The sum of a number and 12.*

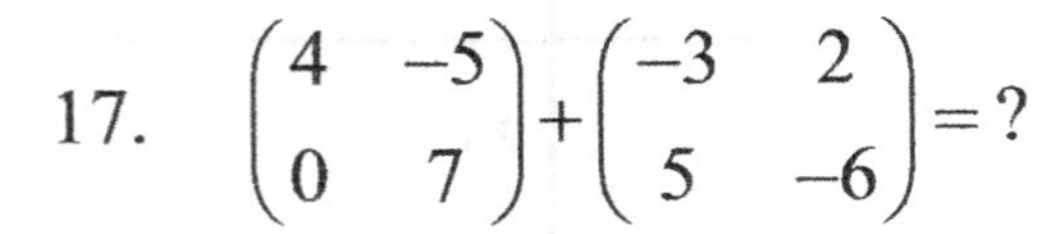

17. $\begin{pmatrix} 4 & -5 \\ 0 & 7 \end{pmatrix} + \begin{pmatrix} -3 & 2 \\ 5 & -6 \end{pmatrix} = ?$
18. Solve for a. $2a + 9 = 6a - 15$
19. What number is $\frac{3}{5}$ of 60?
20. $\frac{4}{9} \times \frac{12}{16} = ?$

1.	2.	3.	4.
5.	6.	7.	8.
9.	10.	11.	12.
13.	14.	15.	16.
17.	18.	19.	20.

Lesson #12

1. Solve for x. $x - 41 = -98$

2. $-19 - (-19) = ?$

3. Evaluate x^y when $x = 5$ and $y = 3$.

4. $-6(-5) = ?$

5. 80% of what number is 15? Round the answer to the nearest tenth.

6. Solve for a. $a + 52 = 90$

7. Find the GCF of $9a^4b^2$ and $21a^2b^3$.

8. $3[24 - (8 + 3 \cdot 2)] = ?$

9. Find the surface area of this cube.

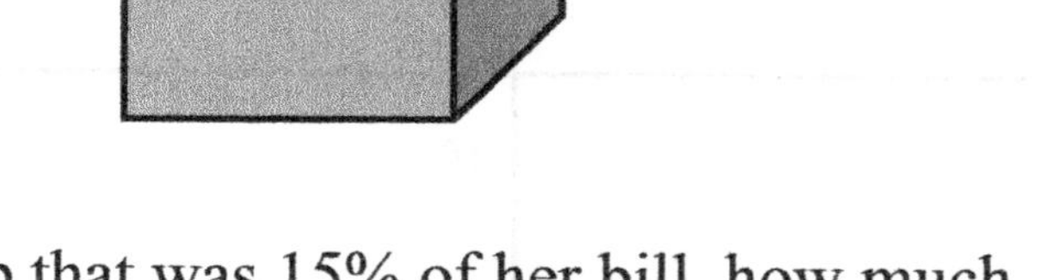

10. Solve for x. $9x = 135$

11. Sue's meal cost $35. If she left a tip that was 15% of her bill, how much was the tip?

12. Simplify. $7(3a - 4b + 8) + 8a + 10$

13. What is the value of a? $\frac{a}{6} = 15$

14. Write *three times a number, decreased by 9* using algebraic symbols.

15. Solve for x. $x + 36 = 122$

16. $\begin{pmatrix} -8 & 4 & 2 \\ -3 & 1 & -7 \end{pmatrix} + \begin{pmatrix} -1 & 3 & -5 \\ 0 & 9 & -6 \end{pmatrix} = ?$

17. $7{,}080 - 3{,}482 = ?$

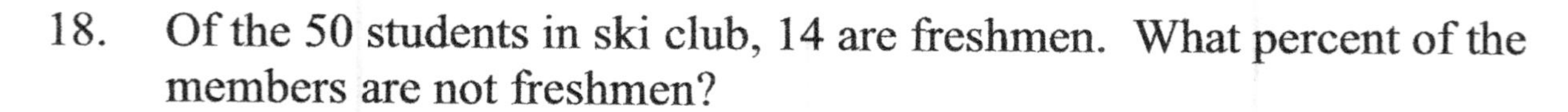

18. Of the 50 students in ski club, 14 are freshmen. What percent of the members are not freshmen?

19. Find the circumference of a circle whose diameter is 20 mm.

20. What is the value of x? $2x + 3x - 5 = -15$

1.	2.	3.	4.
5.	6.	7.	8.
9.	10.	11.	12.
13.	14.	15.	16.
17.	18.	19.	20.

Lesson #13

1. $16 + (-24) + 12 = ?$

2. $7(4x + 5y + 3z - 9) = ?$

3. $\frac{6}{10} \times \frac{8}{12} = ?$

4. Put these integers in increasing order. 38, –16, –1, 28

5. $-76 - (-36) = ?$

6. Round 68,455,203 to the nearest million.

7. Find the area of a circle with a radius of 4 inches.

8. Solve for a. $a - 13 = -31$

9. $72 \div 8 \cdot 3 + 3 \cdot 2^2 = ?$

10. $5\frac{1}{2} + 3\frac{2}{3} = ?$

11. Write $\frac{5}{8}$ as a decimal.

12. Find the value of b. $b + 29 = 64$

13. If $x = 2$ and $y = 3$, what is the value of $8xy - 3y$?

14. Solve for x. $2x - 4 = -16$

15. Write the first 4 prime numbers.

16. What is the value of a? $5a - 4 = 6a + 12$

17. $36.63 \div 0.03 = ?$

18. $2\frac{1}{5} - 1\frac{4}{5} = ?$

Give the coordinates of each point.

19. A ________ C ________

20. E ________ F ________

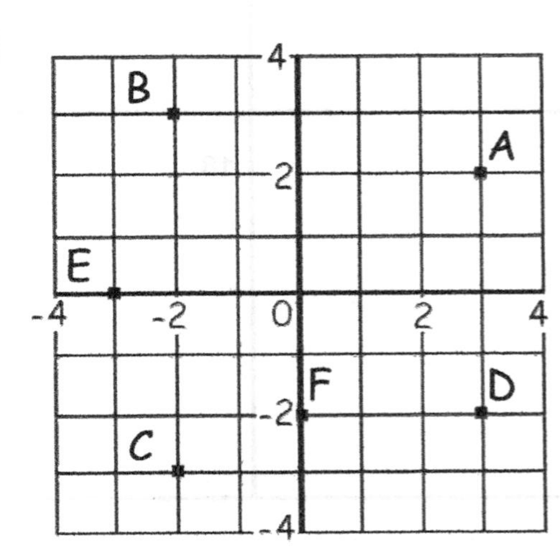

1.	2.	3.	4.
5.	6.	7.	8.
9.	10.	11.	12.
13.	14.	15.	16.
17.	18.	19.	20.

Lesson #14

1. Solve for y. $y + 23 = 47$

2. Write 0.25 as a reduced fraction and as a percent.

3. $3[6 + 2(4 + 2) - 5] + 2^2 = ?$

4. Solve the proportion for x. $\frac{x}{5} = \frac{60}{75}$

5. $-3(8)(-2) = ?$

6. $\frac{120}{-6} = ?$

7. Find the value of x. $3x + 8 = 9x - 16$

8. Solve for x. $\frac{x}{12} = -6$

9. The area of a square is 64 square inches. How long is each side?

10. Solve for a. $a - 9 = -27$

11. $96 - (-58) = ?$

12. What is the value of x? $7x = 98$

13. $\begin{pmatrix} 9 & -6 \\ -1 & 3 \end{pmatrix} - \begin{pmatrix} -3 & 0 \\ 8 & -4 \end{pmatrix} = ?$

14. Simplify. $5(5a + 3b - 6c + 7) + 4(2a - 5c)$

15. A pair of shoes has an original price of \$60. What is the sales price after a 15% discount?

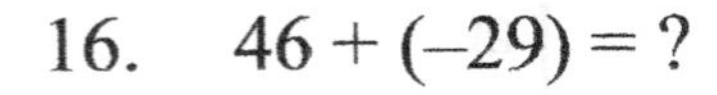

16. $46 + (-29) = ?$

17. $625{,}816 + 499{,}322 = ?$

18. Find the area of this trapezoid.

11 m

7 m

9 m

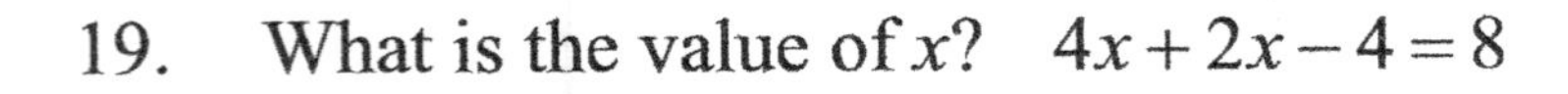

19. What is the value of x? $4x + 2x - 4 = 8$

20. $0.008 \times 0.006 = ?$

1.	2.	3.	4.
5.	6.	7.	8.
9.	10.	11.	12.
13.	14.	15.	16.
17.	18.	19.	20.

Lesson #15

1. Solve for b. $b+39=-80$

2. What percent of 50 is 45?

3. $75+(-26)=?$

4. $-|31|=?$

5. Find the value of a. $\frac{a}{8}=16$

6. $7(7x+6y-9)+4(8x-4)=?$

7. Find the area of a triangle with a base of 17 mm and a height of 4 mm.

8. $\frac{9}{10} \bigcirc \frac{8}{11}$

9. What is the value of x? $x-14=66$

10. Solve for x. $5x-3x+6=-12$

11. $-98-(-67)=?$

12. $\sqrt{196}=?$

13. What value of x makes the fractions equivalent? $\frac{x}{9}=\frac{4}{6}$

14. Calculate the perimeter of the rectangle.

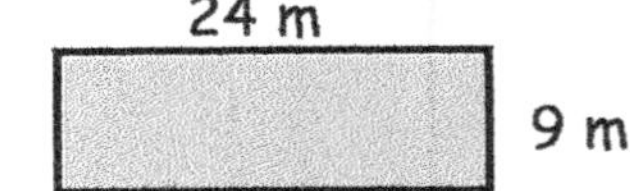

15. Solve for x. $3x=69$

16. What is the P(H, H, T, T,) on four flips of a coin?

17. Write 36% as a decimal and as a reduced fraction.

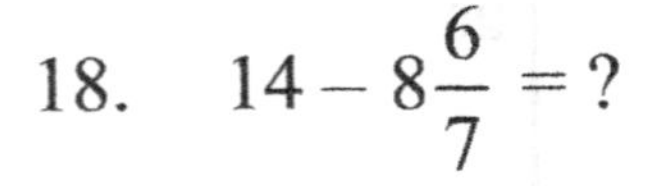

18. $14-8\frac{6}{7}=?$

Identify the point with each set of coordinates.

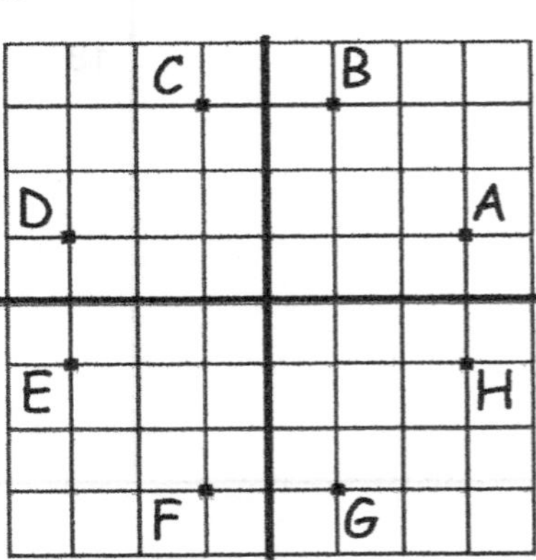

19. (3, –1) ________ (1, 3) ________

20. (–1, –3) ________ (1, –3) ________

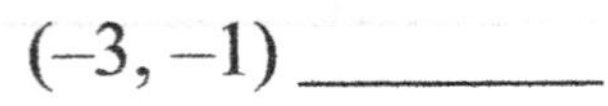

(–3, –1) ________

1.	2.	3.	4.
5.	6.	7.	8.
9.	10.	11.	12.
13.	14.	15.	16.
17.	18.	19.	20.

Lesson #16

1. Solve for b. $b+14=56$
2. $16\frac{1}{2}+8\frac{1}{8}=?$
3. $44-(-28)=?$
4. Simplify. $6(3a+7b-8)+3(3b+2)$
5. The gym teacher keeps 15 wooden bats and 9 metal bats in the equipment room. If a student chooses a bat at random, what is the probability that the bat is metal?
6. Find the value of a. $a-52=112$
7. Evaluate $\frac{5ab}{b}+6a$ when $a=2$ and $b=3$.
8. $62+(-36)=?$
9. $2[6+3(5+3)-10]=?$
10. What is the volume of a rectangular prism if its length is 10 feet, its width is 5 feet, and its height is 3 feet?
11. Simplify. $\frac{12a^2b^4c}{18ab^5c^3}$
12. Solve for x. $\frac{x}{6}=25$
13. $7{,}655 \times 7=?$
14. Put these integers in decreasing order. –26, 0, –9, –17, –36
15. What is the value of x? $7x=105$
16. Round 36.475 to the nearest hundredth.
17. $\begin{pmatrix}-4 & 6 & -9\\ -8 & 2 & -4\end{pmatrix}+\begin{pmatrix}-10 & 5 & 0\\ 6 & -1 & 7\end{pmatrix}=?$
18. Write $\frac{7}{8}$ as a decimal.
19. $90{,}000-46{,}236=?$
20. $-23+(-16)+18=?$

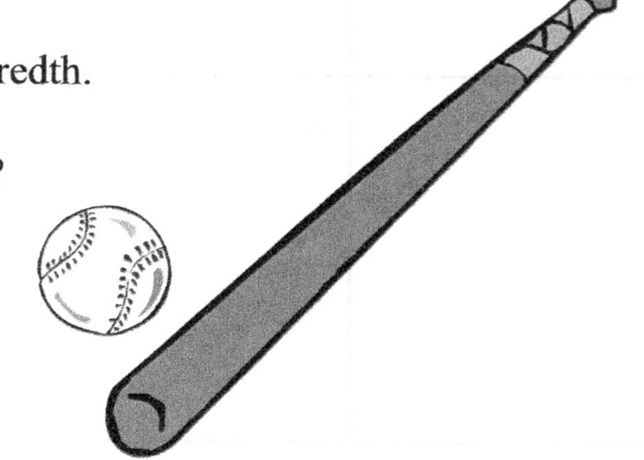

1.	2.	3.	4.
5.	6.	7.	8.
9.	10.	11.	12.
13.	14.	15.	16.
17.	18.	19.	20.

Lesson #17

1. How many yards are in a mile?
2. Solve for a. $a+39=-88$
3. $62 \times 57 = ?$
4. Find the value of x. $9x=108$
5. Write 0.55 as a reduced fraction and as a percent.
6. $378{,}296 + 489{,}244 = ?$

7. $105-(-76)=?$
8. $-6^3 = ?$
9. What is the value of x? $\frac{x}{4}=-18$
10. Tricia is putting water into her aquarium using a pitcher which holds 2 quarts of fluid. Approximately how many pitchers of water will it take so that Tricia will have at least 17 gallons of water in the aquarium?
11. Solve for b. $b-27=-54$
12. Find the circumference of a circle with a diameter of 16 mm.
13. Solve the proportion for x. $\frac{9}{12}=\frac{x}{156}$
14. $\frac{8}{10} \div \frac{4}{10} = ?$
15. A triangle with no congruent sides is a(n) __________ triangle.
16. $6.2 + 49.763 = ?$
17. $\frac{3}{12} \cdot \frac{6}{9} = ?$
18. Distribute and combine like terms. $3(9a+8b+7)+4(5a+9)$
19. What is the value of a? $9a-2=15a$
20. Solve for x. $x-17=-60$

1.	2.	3.	4.
5.	6.	7.	8.
9.	10.	11.	12.
13.	14.	15.	16.
17.	18.	19.	20.

Lesson #18

1. $-13 + (-27) = ?$

2. Justin earns $54 mowing 3 lawns. If he charges the same amount for each lawn, how much will he earn for mowing 5 lawns?

3. Solve for x. $-12x = 144$

4. What is the range of Beth's test scores?

Beth's Test Scores

Stem	Leaf
5	9
6	8
7	5
8	3 7 9 9
9	0 4

5. $-71 - (-71) = ?$

6. $0.6 - 0.375 = ?$

7. $5[3(4 + 3) - 10] + 3^2 = ?$

8. Find the median and the mode of 23, 6, 19, 56, 88, 77 and 23.

9. Distribute and combine terms. $7(3s + 4t + 5) + 8(5t + 6)$

10. Put the measurements in order from largest to smallest volume.

    Cup Quart Gallon Pint

11. Solve for x. $x + 19 = -49$

12. Write 36% as a decimal and as a reduced fraction.

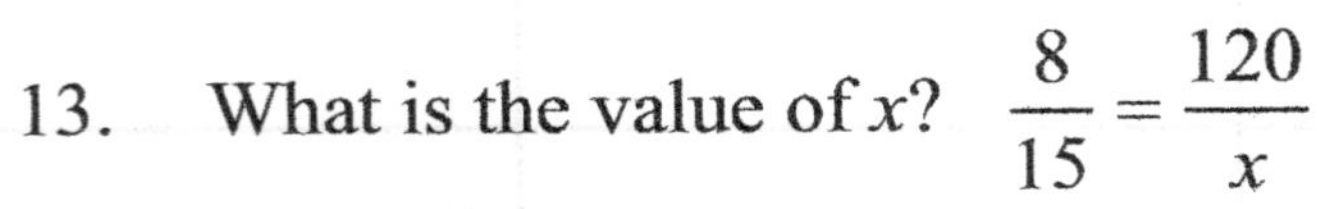

13. What is the value of x? $\dfrac{8}{15} = \dfrac{120}{x}$

14. Find $\dfrac{4}{5}$ of 80.

15. Solve for x. $3x - 15 = -45$

16. $\dfrac{-150}{-15} = ?$

17. When $a = 4$ and $b = 2$, what is the value of $7ab + 2b - 8$?

18. Write an algebraic phrase for *the product of 3 and a number, and 15.*

19. Find the value of x. $\dfrac{8}{9}x - \dfrac{7}{9}x + 4 = 12$

20. Simplify. $\dfrac{14a^2bc^4}{21abc^6}$

1.	2.	3.	4.
5.	6.	7.	8.
9.	10.	11.	12.
13.	14.	15.	16.
17.	18.	19.	20.

Lesson #19

1. $86+(-29)=?$

2. Write 0.16 as a reduced fraction and as a percent.

3. Solve for a. $a-12=-43$

4. $-7(-4)(3)=?$

5. $\begin{pmatrix} 3 & 0 \\ -1 & 6 \end{pmatrix} + \begin{pmatrix} -9 & 8 \\ 3 & -7 \end{pmatrix} = ?$

6. When $a=2$ and $b=3$, find the value of $4ab-2a+5$.

7. Solve for x. $3x-5=16$

8. Find the perimeter of the regular hexagon.

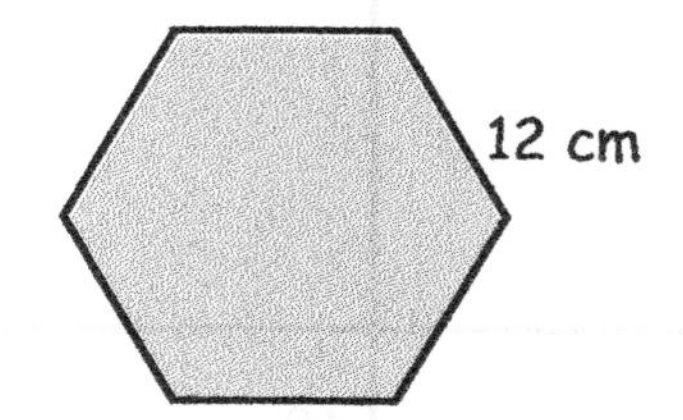

9. Find the value of x. $\frac{x}{9}=13$

10. Mr. Jones drinks 10 soft drinks in 5 days. At that rate, how many soft drinks will he drink in 60 days?

11. $3(12\div 4)+6-3=?$

12. $5(2+4)+15\div(9-6)=?$

13. $-76-(-23)=?$

14. Which is greater, 0.62 or $\frac{3}{5}$?

15. Find the area of a parallelogram whose base is 22 yards and whose height is 5 yards.

16. 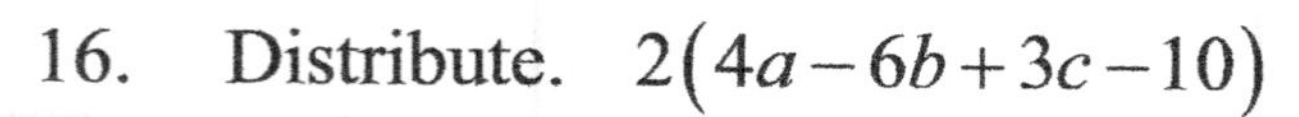 Distribute. $2(4a-6b+3c-10)$

17. Solve for x. $\frac{3}{8}=\frac{x}{112}$

18. $\frac{-50}{5}=?$

19. $4{,}899{,}766+7{,}346{,}092=?$

20. A child's body weight is approximately 75% water. About how many pounds of an 80 pound child's weight is water?

1.	2.	3.	4.
5.	6.	7.	8.
9.	10.	11.	12.
13.	14.	15.	16.
17.	18.	19.	20.

Lesson #20

1. Solve for a. $a+16=-89$

2. Write $\frac{4}{5}$ as a decimal and as a percent.

3. $[7+3\cdot2+8]\div7=?$

4. Find the area of a triangle if its base measures 20 inches and its height is 4 inches.

5. What were the total points scored by the Knights?

Point Scored by Knights

Stem	Leaf
0	6 7 7 9
1	0 2 4
2	1 1 7

6. Yolanda's scores in the diving meet were 7.8, 8.4, 8.1, 7.6, 8.4, 8.3 and 7.9. What are the mode and the median of her scores?

7. $-|-31|=?$

8. $34\times26=?$

9. $8{,}060-3{,}365=?$

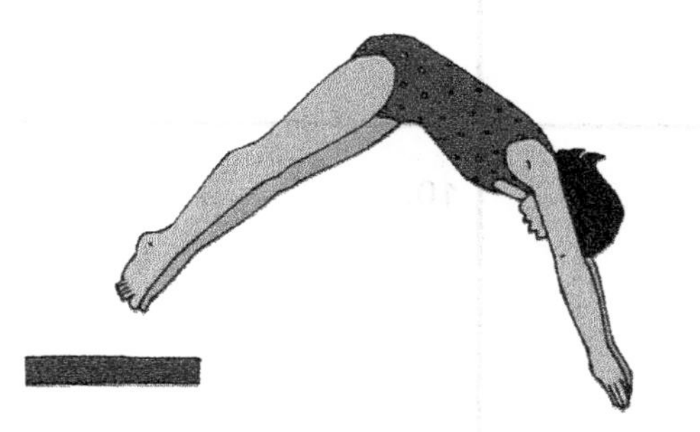

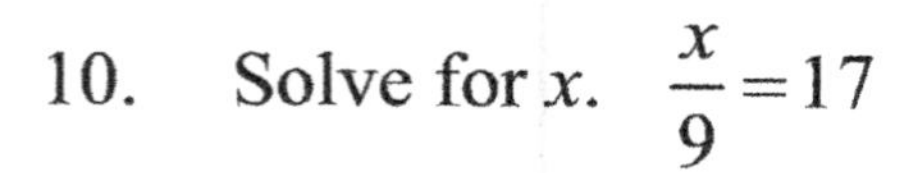

10. Solve for x. $\frac{x}{9}=17$

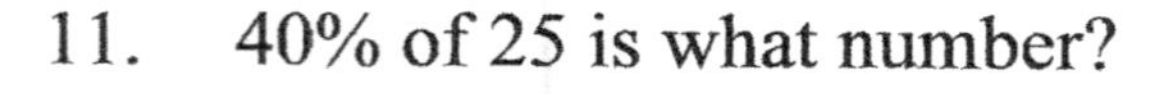

11. 40% of 25 is what number?

12. What is the value of x? $3x+8=2$

13. When $a=6$, $b=5$ and $c=3$, what is the value of $\frac{ab}{2}+4c$?

14. Find the value of x. $\frac{x}{3}+17=21$

15. Solve for b. $b-25=-75$

16. $15\frac{2}{5}+10\frac{1}{4}=?$

17. Simplify. $\sqrt{225}-\sqrt{100}+3^3$

18. $(-5)^4=?$

19. An angle that measures greater than 90° is a(n) _____________ angle.

20. $0.15\times3.5=?$

1.	2.	3.	4.
5.	6.	7.	8.
9.	10.	11.	12.
13.	14.	15.	16.
17.	18.	19.	20.

Lesson #21

1. Solve for y. $y-12=27$

2. Simplify. $\dfrac{9x^2y^3z}{12xyz}$

3. $56-(-38)=?$

4. 40% of what number is 12?

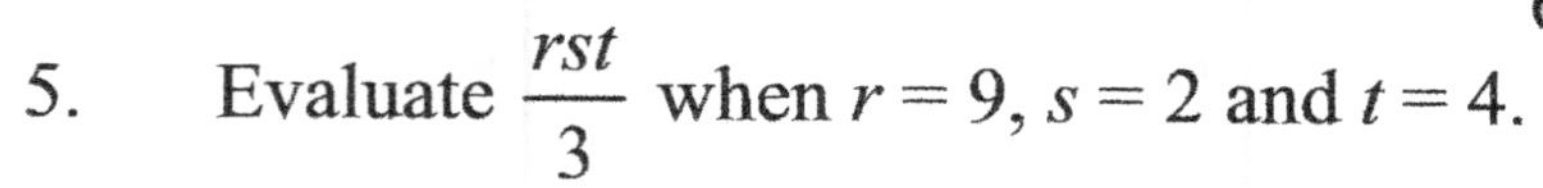

5. Evaluate $\dfrac{rst}{3}$ when $r=9$, $s=2$ and $t=4$.

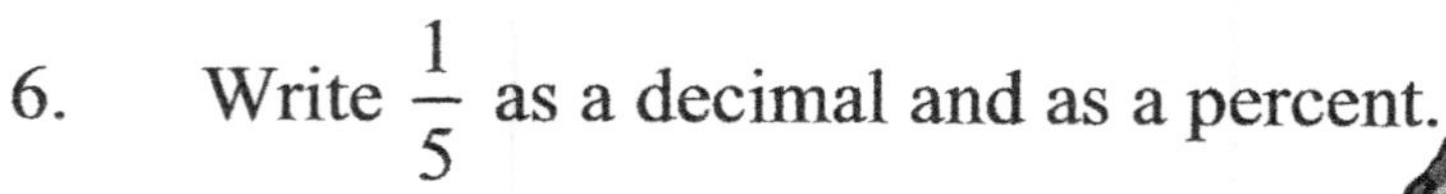

6. Write $\dfrac{1}{5}$ as a decimal and as a percent.

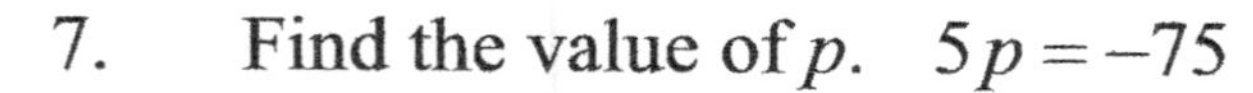

7. Find the value of p. $5p=-75$

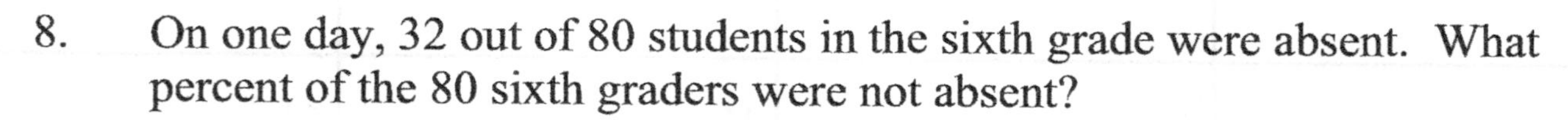

8. On one day, 32 out of 80 students in the sixth grade were absent. What percent of the 80 sixth graders were not absent?

9. What is the value of x? $3x+x+3=-13$

10. Find $\dfrac{2}{3}$ of 21.

11. Distribute and simplify. $5(5a-7b+6)+4(3a-4b)$

12. $-44+(-39)=?$

13. Solve for b. $b+13=-30$

14. Find the circumference of a circle if its diameter is 13 inches.

15. $-8(3)(-2)=?$

16. Solve for x. $\dfrac{x}{12}=14$

17. Round 92,466,502 to the nearest ten thousand.

18. Solve the proportion for x. $\dfrac{5}{12}=\dfrac{x}{180}$

19. $\begin{pmatrix} 3 & 5 \\ -4 & 2 \end{pmatrix}+\begin{pmatrix} 6 & -2 \\ 8 & 5 \end{pmatrix}=?$

20. Solve for x. $\dfrac{x}{5}+4=-16$

1.	2.	3.	4.
5.	6.	7.	8.
9.	10.	11.	12.
13.	14.	15.	16.
17.	18.	19.	20.

Lesson #22

1. $99 + (-44) = ?$

2. $-|49| = ?$

3. Translate *three times a number, divided by 2* into an algebraic phrase.

4. $4[6 + 4(3 + 2) + 3^2] = ?$

5. $425{,}363 + 878{,}666 = ?$

6. If $x = 4$ and $y = 2$, what is the value of $5xy + 2x + 8$?

7. How many inches are in 4 yards?

8. $4 - 2\frac{7}{8} = ?$

9. $17 - (-12) = ?$

10. Find the GCF of $16x^2yz^3$ and $24xy^2z^4$.

11. $6{,}432 - 1{,}799 = ?$

12. $86 + (-24) + 16 = ?$

13. $\frac{8}{10} \div \frac{2}{5} = ?$

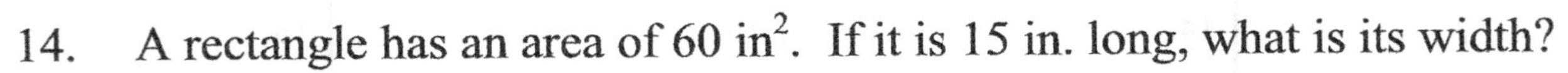

14. A rectangle has an area of 60 in^2. If it is 15 in. long, what is its width?

15. **A function is a relationship where each member of the domain is paired with exactly one member of the range.** Is this relation a function?

x	y
2	-4
0	2
-1	1

16. $\frac{4}{5} \times \frac{10}{16} = ?$

17. Write 36% as a decimal and as a reduced fraction.

18. $3.7 \times 2.4 = ?$

19. Order these integers from greatest to least. $-19,\ -14,\ -2,\ -21$

20. $\begin{pmatrix} -6 & -1 & 7 \\ 3 & -2 & -5 \end{pmatrix} - \begin{pmatrix} -8 & 6 & 2 \\ 14 & -3 & 1 \end{pmatrix} = ?$

1.	2.	3.	4.
5.	6.	7.	8.
9.	10.	11.	12.
13.	14.	15.	16.
17.	18.	19.	20.

Lesson #23

1. What is the value of a? $a+27=-35$

2. Find the value of $\frac{6a}{b}+3c$ when $a=4$, $b=2$ and $c=3$.

3. $\frac{9}{16}\times\frac{4}{18}=?$

4. Write 0.18 as a reduced fraction and as a percent.

5. $\frac{-120}{6}=?$

6. $3.2-1.892=?$

7. Solve for x. $5x-5=-15$

8. Find the value of x. $\frac{8}{15}=\frac{96}{x}$

9. $62+(-37)=?$

10. Simplify. $\frac{14a^3b^2c}{21abc^3}$

11. Solve for x. $x-15=-46$

12. Determine the perimeter of a square if a side measures 18 mm.

13. In a survey of 4,000 people, 45% said that they had visited an amusement park during the past year. How many people said they visited an amusement park?

14. $5^2-2[5(5+4)-25]=?$

15. A triangle with 2 congruent sides is a(n) __________ triangle.

16. $-105-(-88)=?$

17. $7\frac{2}{3}+3\frac{1}{5}=?$

18. What is the area of the parallelogram?

5 cm

20 cm

19. $0.005\times0.04=?$

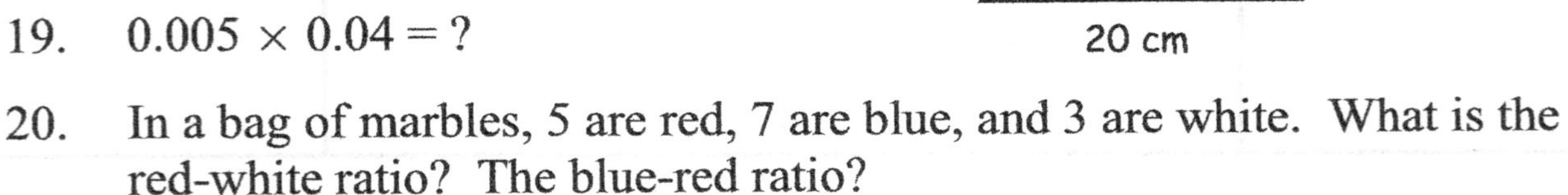

20. In a bag of marbles, 5 are red, 7 are blue, and 3 are white. What is the red-white ratio? The blue-red ratio?

1.	2.	3.	4.
5.	6.	7.	8.
9.	10.	11.	12.
13.	14.	15.	16.
17.	18.	19.	20.

Lesson #24

1. $93 + (-58) = ?$

2. $0.92 - 0.7346 = ?$

3. How many ounces are in 6 pounds?

4. Solve for x. $\frac{1}{5}x - 3 = -2$

5. When $a = 2$ and $b = 3$, what is the value of $\frac{5ab}{2b} + 6$?

6. What is the value of x? $x + 15 = -45$

7. $6\frac{2}{3} + 4\frac{1}{8} = ?$

8. Write 90% as a decimal and as a reduced fraction.

9. Solve for y. $5y + 2(y - 3) = 92$

10. $66 - (-32) = ?$

11. $2[6 + 3(4 + 2)] + 3^2 = ?$

12. Solve for a. $7a + 4 = 5a + 8$

13. $-|12| = ?$

14. A serving of snack mix is $\frac{2}{3}$ cup. If you want to take 15 servings to your neighbor's party, how many cups of snack mix should you take?

15. Distribute and combine terms. $7(3x - 4y + 7) + 5(2x - 3y + 5)$

16. How many decades are 80 years?

17. Solve for x. $\frac{x}{7} = 14$

18. $-90 + (-41) = ?$

19. Round 36,412,006 to the nearest thousand.

20. Find the missing numbers for the function, $2x - 1 = y$.

x	y
2	?
0	?
-2	?

1.	2.	3.	4.
5.	6.	7.	8.
9.	10.	11.	12.
13.	14.	15.	16.
17.	18.	19.	20.

Lesson #25

1. Find the value of a. $a+15=-80$

2. Your food bill at a restaurant is \$60. The sales tax is 7%, and you leave a 20% tip. What is the total cost of your meal? (Hint: Both the sales tax and the tip are percentages of the original bill.)

3. Solve for x. $\frac{x}{5}=12$

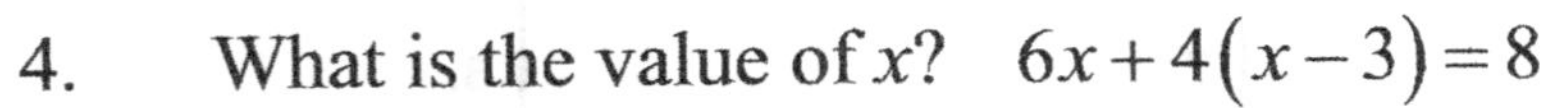

4. What is the value of x? $6x+4(x-3)=8$

5. Write 0.8 as a reduced fraction.

6. $[1+3(9+12)]-4^2=?$

7. Find $\frac{3}{5}$ of 80.

8. Solve for b. $b-12=-46$

9. $-7(3)(-2)=?$

10. Solve for x. $\frac{1}{8}x-14=6$

11. How many centimeters are in 8 meters?

12. Write an algebraic expression for *eight times a number, divided by 2.*

13. What value of x makes the fractions equivalent? $\frac{9}{5}=\frac{135}{x}$

14. Solve for x. $12x=144$

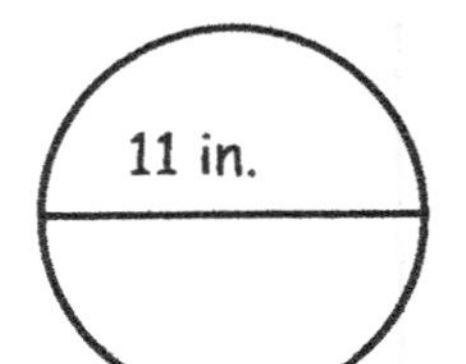

15. Find the value of y. $2y-12=4$

16. Determine the circumference of the circle.

17. Distribute and simplify. $5(6a-3b+7c)+3(5a-3b+6)$

18. $\frac{-44}{4}=?$

19. Find the GCF of $15a^3bc^2$ and $25abc^2$.

20. $\begin{pmatrix} 4 & 3 & 6 \\ -1 & 5 & -7 \end{pmatrix}+\begin{pmatrix} 3 & -2 & 5 \\ -6 & 4 & -9 \end{pmatrix}=?$

1.	2.	3.	4.
5.	6.	7.	8.
9.	10.	11.	12.
13.	14.	15.	16.
17.	18.	19.	20.

Lesson #26

1. $19 + (-8) = ?$

2. Solve for x. $\frac{1}{3}x + 7 = 25$

3. $3{,}476{,}212 + 7{,}562{,}325 = ?$

4. Write 56% as a decimal and as a reduced fraction.

5. $-50 - (-28) = ?$

6. Round 86,421,116 to the nearest million.

7. Simplify. $8(4x - 5y - 9)$

8. $8\frac{1}{2} - 3\frac{3}{4} = ?$

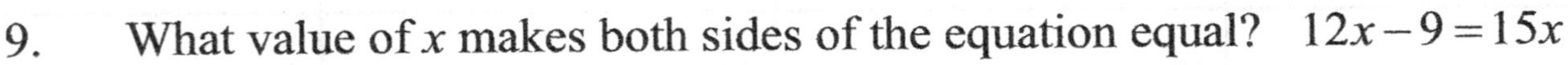

9. What value of x makes both sides of the equation equal? $12x - 9 = 15x$

10. Solve for x. $x - 9 = 28$

11. $38 \times 51 = ?$

12. $45 \div 5 \cdot 2 + 3 \cdot 7 - 5 = ?$

13. What is the area of a parallelogram with a 16 inch base and a height that is 7 inches?

14. Find the missing numbers in the function table for $y = 3x$.

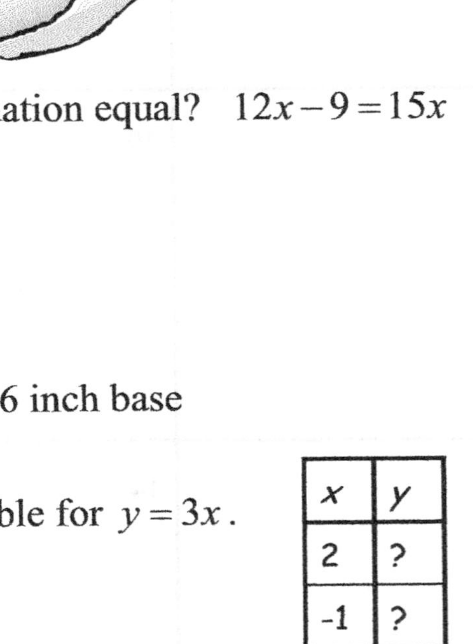

x	y
2	?
-1	?
?	15

15. $\frac{5}{9} + \frac{2}{3} = ?$

16. $2(-9)(-2) = ?$

17. Solve for x. $2x - 10 = -30$

18. Find the value of b. $b + 13 = -50$

19. In the chart, the measure of the angle formed by the Wheat section is 150° and the angle formed by the Corn section is 75°. What is the measure of the Beans section?

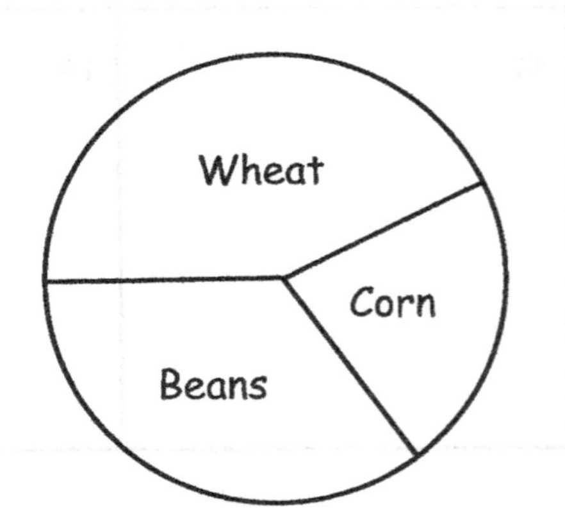

20. $\frac{-45}{-9} = ?$

1.	2.	3.	4.
5.	6.	7.	8.
9.	10.	11.	12.
13.	14.	15.	16.
17.	18.	19.	20.

Lesson #27

1. $-50 + (-29) = ?$

2. Solve for x. $\frac{4}{5}x = 20$

3. $\frac{-14}{2} = ?$

4. $81 \div 9 \cdot 6 - 14 \div 2 = ?$

5. Find the value of n. $4n + 2n - 6 = 12$

6. $-6(8) = ?$

7. Factor the expression. $8a^3b^2$

8. Evaluate $5ab - 2b$ when $a = 3$ and $b = 4$.

9. Solve for x. $\frac{1}{5}x - 10 = -15$

10. $5{,}508 \div 54 = ?$

11. $112 - (-64) = ?$

12. What is the value of x? $x + 19 = 36$

13. Calculate the surface area of the prism.

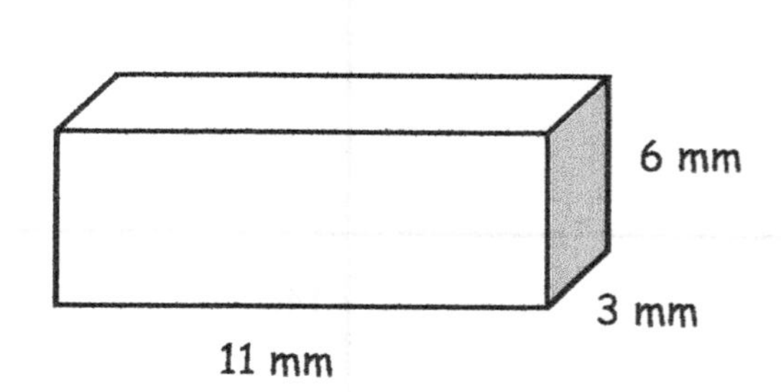

14. $15 - 9\frac{2}{9} = ?$

15. Write $\frac{2}{25}$ as a decimal.

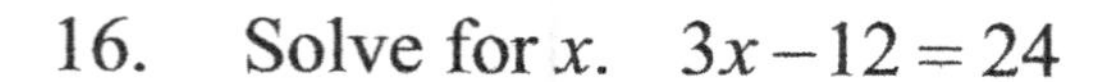

16. Solve for x. $3x - 12 = 24$

17. Find the value of x. $\frac{5}{8} = \frac{x}{104}$

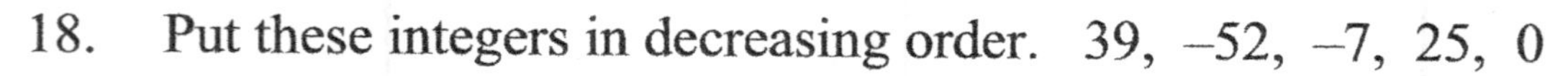

18. Put these integers in decreasing order. 39, –52, –7, 25, 0

19. $75{,}000 - 29{,}561 = ?$

20. $\begin{pmatrix} 10 & -4 \\ 8 & 6 \end{pmatrix} - \begin{pmatrix} 8 & -2 \\ 7 & -3 \end{pmatrix} = ?$

1.	2.	3.	4.
5.	6.	7.	8.
9.	10.	11.	12.
13.	14.	15.	16.
17.	18.	19.	20.

Lesson #28

1. $375 + 98 + 26 = ?$

2. $\dfrac{88}{-11} = ?$

3. $365 \times 22 = ?$

4. What is the P(3, 1, 2) on 3 rolls of a die?

5. Solve for b. $b + 51 = -88$

6. $42 \div 6 + 7 - 4 + 2^3 = ?$

7. Calculate the area of the trapezoid.

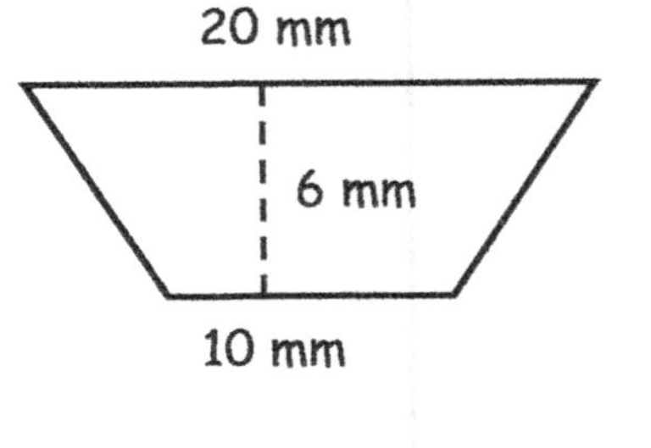

8. $91 - (-91) = ?$

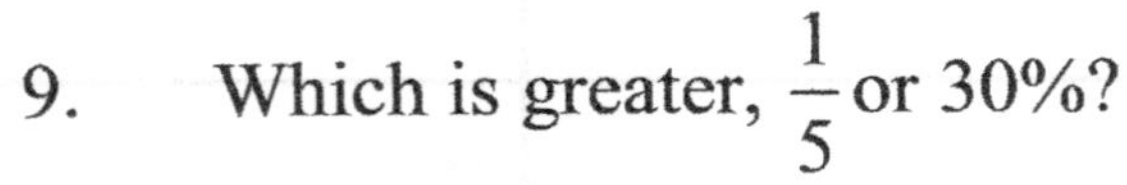

9. Which is greater, $\frac{1}{5}$ or 30%?

10. If $a = 2$, $b = 4$ and $c = 5$, what is the value of $\dfrac{abc}{b} + a^2$?

11. Solve for x. $\dfrac{x}{8} = 26$

12. Simplify. $9(5x + 3y - 9) + 2(4x - 8)$

13. Find the value of a. $5a + 2a + 14 = -7$

14. Write an algebraic expression for *the sum of a number and 17.*

15. Solve for a. $a - 17 = -27$

16. $132 + (-65) = ?$

17. What is the value of a? $8a = 152$

18. Solve for a. $\frac{1}{7}a - 6 = -12$

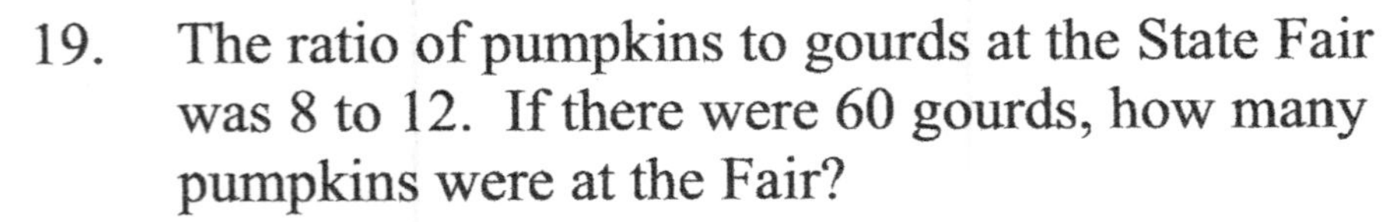

19. The ratio of pumpkins to gourds at the State Fair was 8 to 12. If there were 60 gourds, how many pumpkins were at the Fair?

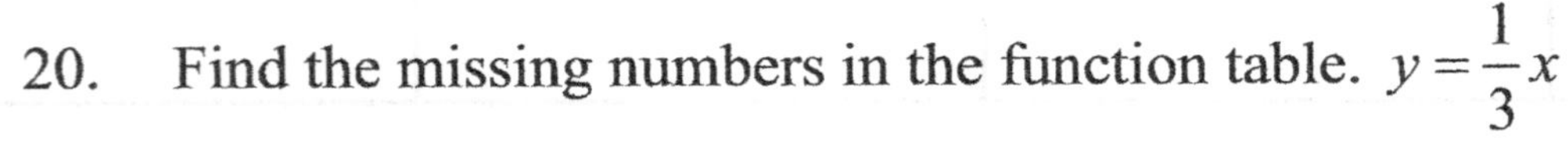

20. Find the missing numbers in the function table. $y = \frac{1}{3}x$

x	y
12	?
9	?
?	6

1.	2.	3.	4.
5.	6.	7.	8.
9.	10.	11.	12.
13.	14.	15.	16.
17.	18.	19.	20.

Lesson #29

1. $13-(-7)=?$
2. Round 4,651,007 to the nearest hundred thousand.
3. Solve for a. $5a+3=-12$
4. What number is 60% of 25?
5. Find the value of x. $\frac{1}{5}x-12=22$
6. Which is greater, 0.65 or $\frac{7}{20}$?
7. How many cups are in 4 pints?
8. $2[7+2(5+1)-3]=?$
9. What is the value of a? $7a-8=15a+8$

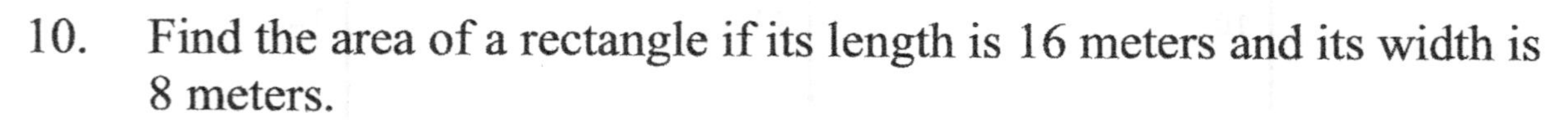

10. Find the area of a rectangle if its length is 16 meters and its width is 8 meters.
11. $8\frac{1}{6}+3\frac{2}{5}=?$
12. Factor the expression. $12p^4q$
13. $\frac{8}{12}\times\frac{10}{16}=?$
14. Find the value of $xyz+2y$ when $x=2$, $y=3$ and $z=5$.
15. Solve for a. $15a=75$
16. A board $2\frac{3}{4}$ inches thick is placed on top of a board $\frac{5}{6}$ inch thick. What is the combined thickness of the two boards?
17. $\frac{-51}{-3}=?$
18. Solve for x. $\frac{x}{10}=-14$
19. $-52+(-18)=?$
20. Find the missing values in the function table. $y=2x+1$

x	y
4	?
7	?
0	?

1.	2.	3.	4.
5.	6.	7.	8.
9.	10.	11.	12.
13.	14.	15.	16.
17.	18.	19.	20.

Lesson #30

1. What is the value of a? $7a-5=23$
2. Solve for x. $\frac{x}{12}=13$
3. $\begin{pmatrix} 4 & 8 \\ -3 & 6 \end{pmatrix} - \begin{pmatrix} 2 & -3 \\ -5 & 1 \end{pmatrix} = ?$
4. $862{,}444 + 779{,}816 = ?$
5. Find the value of x. $x+17=-38$
6. You have \$20 to spend on potato chips for a party. Potato chips cost \$2.49 a bag, including tax. How many bags of chips can you buy? How much money is left over?
7. What value of x makes the equation true? $5x+7=2x-14$
8. Solve for x. $\frac{1}{6}x+5=13$
9. $4{,}641 \times 4 = ?$
10. $5 + 2[8 \cdot 2 - 6] = ?$
11. What is the value of y? $6y+2y-8=16$
12. How many pounds are in 5 tons?
13. Factor the expression. $50s^2t^5$
14. Find $\frac{2}{3}$ of 21.
15. $\frac{-364}{4} = ?$
16. $60{,}000 - 28{,}462 = ?$
17. $\frac{8}{12} \div \frac{2}{3} = ?$
18. Solve for x. $\frac{10}{8} = \frac{150}{x}$
19. If $a=3$ and $b=2$, find the value of $\frac{4ab}{2b}$.
20. Find the missing numbers in the function table.

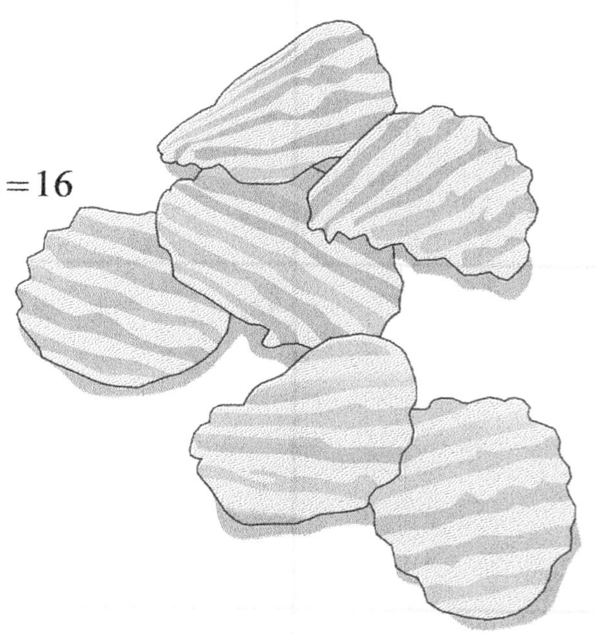

$y = 8 - x$

x	y
7	?
1	?
4	?

1.	2.	3.	4.
5.	6.	7.	8.
9.	10.	11.	12.
13.	14.	15.	16.
17.	18.	19.	20.

Lesson #31

1. $-9(10) = ?$

2. $-44 + (25) = ?$

3. Write an algebraic expression that means *a number divided by 4, decreased by 6.*

4. Is π a rational number or an irrational number? (See Help Pages.)

5. What is the percent of change from \$12 to \$9?

6. Write 0.42 as a reduced fraction and as a percent.

7. $\dfrac{-4{,}824}{2} = ?$

8. Is the number 5,225 divisible by 5? By 10? How do you know?

9. Find the LCM of $8a^3b^2$ and $18ab$.

10. Solve for x. $4x - 8 = 16$

11. Factor the expression. $9x^3$

12. $91 + (-65) = ?$

13. Solve for y. $\frac{1}{9}y + 4 = -12$

14. Find the area of a circle with a radius of 7 feet.

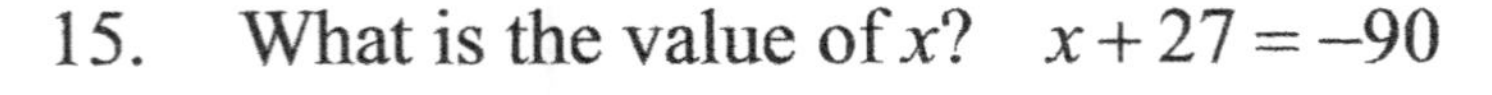

15. What is the value of x? $x + 27 = -90$

16. $16 \div 2 \cdot 3 + 20 \div 5 + 2^2 = ?$

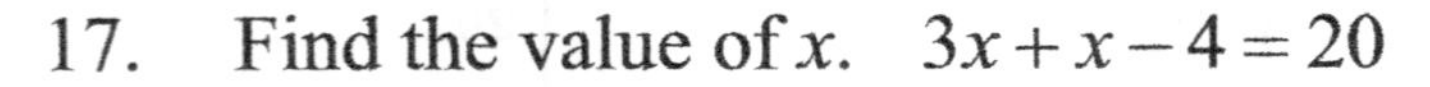

17. Find the value of x. $3x + x - 4 = 20$

18. Solve for x. $\frac{4}{5}x - \frac{3}{5}x + 4 = 11$

19. In the equation, $b - 31 = 60$, what is the value of b?

20. Put these integers in decreasing order. $-22,\ 14,\ -37,\ -1,\ 2$

1.	2.	3.	4.
5.	6.	7.	8.
9.	10.	11.	12.
13.	14.	15.	16.
17.	18.	19.	20.

Lesson #32

1. $13 \times 27 = ?$
2. $5(-3)(4) = ?$
3. Factor the expression. $15x^4y^2$
4. What value of a makes the equation true? $6a - 12 = 42 + 5a$
5. Simplify. $\dfrac{12xy^3z^4}{18x^2y^3z}$
6. How long are the sides of a square whose area is 144 sq. meters?
7. The ingredients of a multiple vitamin tablet include 20% vitamin C, 0.8% vitamin D, and 0.06% zinc. Put the percents in increasing order.
8. $4{,}286 + 8{,}554 = ?$
9. What is the value of $7xy + y$ when $x = 3$ and $y = 2$?
10. $8{,}020 - 3{,}335 = ?$
11. Solve for x. $3x - 8 = 28$
12. Write 44% as a decimal and as a reduced fraction.
13. $\sqrt{1{,}444} = ?$
14. Find $\dfrac{4}{5}$ of 60.
15. Solve for x. $\dfrac{1}{4}x + 8 = -12$
16. Find the value of x. $\dfrac{9}{6} = \dfrac{x}{90}$
17. $2\dfrac{1}{9} + 3\dfrac{3}{5} = ?$
18. Solve for y. $y + 17 = -49$
19. What is the value of x? $\dfrac{x}{12} = 13$
20. Tell whether $\sqrt{25}$ is rational or irrational.

1.	2.	3.	4.
5.	6.	7.	8.
9.	10.	11.	12.
13.	14.	15.	16.
17.	18.	19.	20.

Lesson #33

1. $-32(3) = ?$

2. $13 + (-39) = ?$

3. $\dfrac{525}{-5} = ?$

4. Find the value of b. $b + 16 = -38$

5. Solve for a. $8a - 11 = 4a + 13$

6. Simplify. $8(3a - 5b + 6c - 7)$

7. $4[3 + 2(7 + 5) - 7] = ?$

8. Determine the area of the triangle.

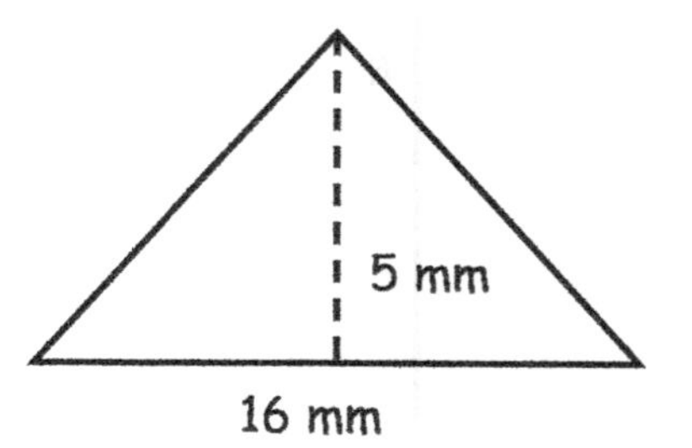

9. What is the value of a? $a - 21 = 101$

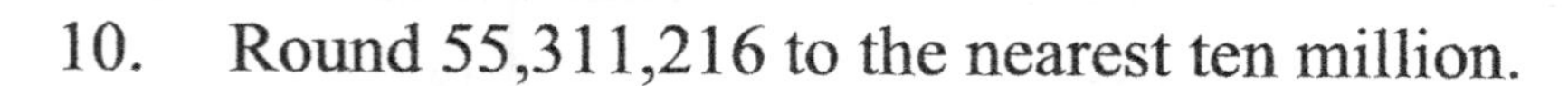

10. Round 55,311,216 to the nearest ten million.

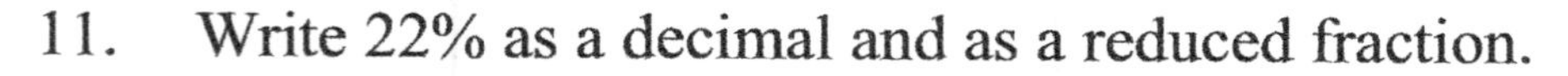

11. Write 22% as a decimal and as a reduced fraction.

12. Rewrite the phrase *seven times a number, increased by 6* using algebraic symbols.

13. Solve for x. $\frac{2}{3}x - \frac{1}{3}x + 4 = 16$

14. Find the value of a. $13a = 169$

15. $6 - 1\frac{2}{7} = ?$

16. Solve for x. $6x + 2x - 9 = 23$

17. $-52 - (25) = ?$

18. What is the value of x? $\dfrac{x}{7} = 14$

19. $\begin{pmatrix} 6 & -3 & 0 \\ -5 & 4 & 8 \end{pmatrix} + \begin{pmatrix} 5 & -2 & 7 \\ -6 & 5 & -3 \end{pmatrix} = ?$

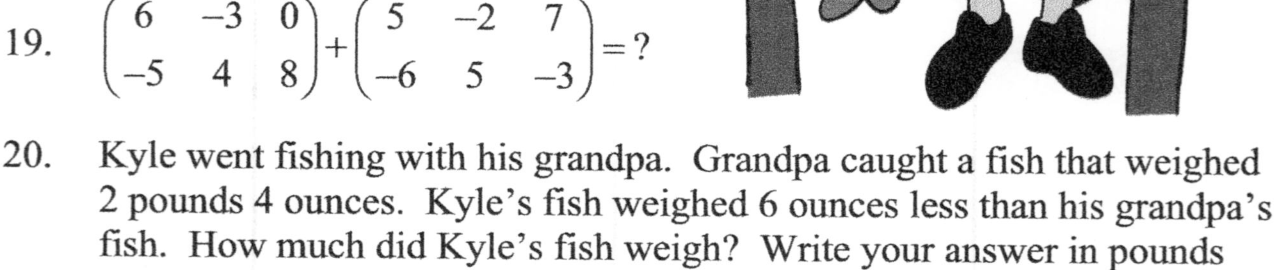

20. Kyle went fishing with his grandpa. Grandpa caught a fish that weighed 2 pounds 4 ounces. Kyle's fish weighed 6 ounces less than his grandpa's fish. How much did Kyle's fish weigh? Write your answer in pounds and ounces.

1.	2.	3.	4.
5.	6.	7.	8.
9.	10.	11.	12.
13.	14.	15.	16.
17.	18.	19.	20.

Lesson #34

1. Find the missing measurement, x.

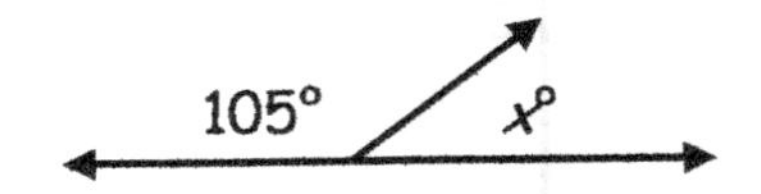

2. $-18 - (-18) = ?$

3. In his first 6 games, Dwayne averaged 10 points per game. In the next 9 games, he averaged 15 points per game. How many points per game did he average during all 15 games?

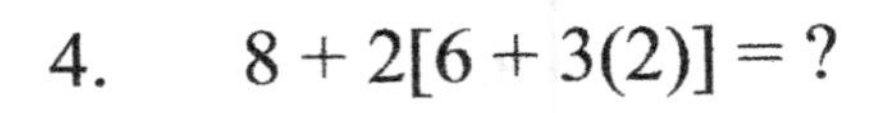

4. $8 + 2[6 + 3(2)] = ?$

5. $36(-2) = ?$

6. Solve for y. $y + 31 = -98$

7. When $a = 2$, $b = 3$ and $c = 4$, evaluate $\frac{abc}{2}$.

8. How many decades are 80 years?

9. $\frac{48}{-8} = ?$

10. Find the median of 16, 49, 86, 28 and 50.

11. $101 + (57) = ?$

12. Find the value of x. $\frac{3}{4}x - \frac{2}{4}x + 2 = 12$

13. Write 0.13 as a fraction and as a percent.

14. $0.008 \times 0.03 = ?$

15. 75% of 60 is what number?

16. $\frac{5}{6} \times \frac{12}{15} = ?$

17. Solve for b. $b - 22 = -50$

18. Round 38.264 to the nearest tenth.

19. $0.8 - 0.362 = ?$

20. Tell whether $\sqrt{7}$ is rational or irrational.

1.	2.	3.	4.
5.	6.	7.	8.
9.	10.	11.	12.
13.	14.	15.	16.
17.	18.	19.	20.

Lesson #35

1. $-7(-8) = ?$

2. Solve for n. $n - 18 = -85$

3. A baseball diamond has 90 feet between each of the four bases. When Lance runs, he travels about one yard for each step. About how many steps does Lance take around the baseball diamond?

4. Simplify. $\dfrac{18a^2bc^4}{24ab^2c}$

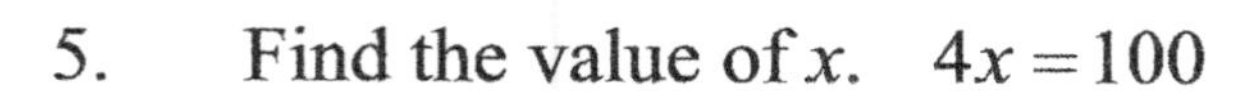

5. Find the value of x. $4x = 100$

6. $108 - (-52) = ?$

7. Round 37,602,314 to the nearest ten thousand.

8. What is the value of x? $4x - 10 = 22$

9. Solve for x. $\dfrac{x}{5} = 24$

10. Find the corresponding y-values in this function table.

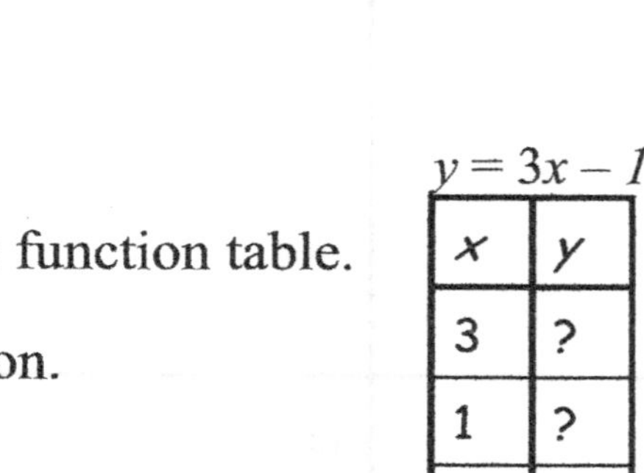

$y = 3x - 1$

x	y
3	?
1	?
0	?

11. Write 67% as a decimal and as a fraction.

12. $133 + (-86) = ?$

13. $4{,}020 - 1{,}789 = ?$

14. $3 \cdot 8 + (4 \cdot 3) - 6 + 5 = ?$

15. How many cups are in 6 pints?

16. $0.0006 \times 0.005 = ?$

17. Find the surface area of this cube.

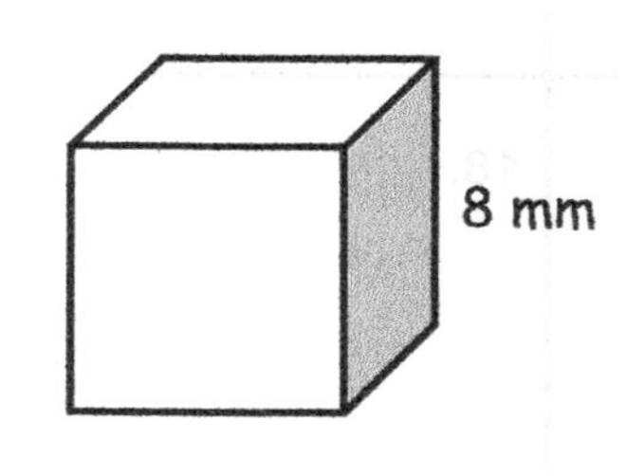

18. Solve for x. $\dfrac{5}{6}x + \dfrac{4}{6}x + 7 = 20$

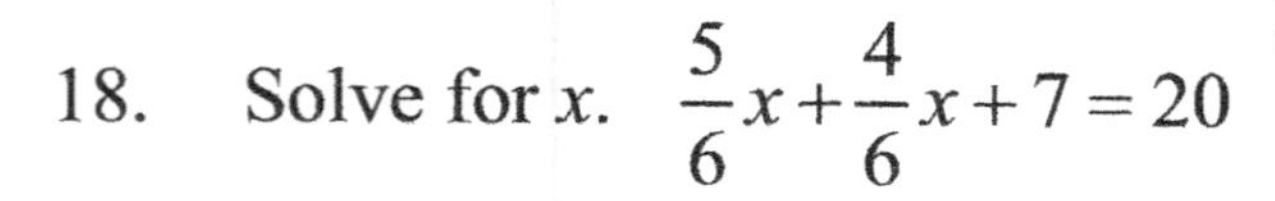

19. $42.9 + 3.75 = ?$

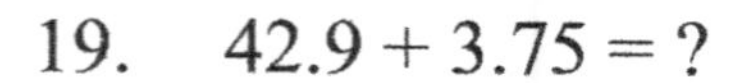

20. Find the missing measurement, x.

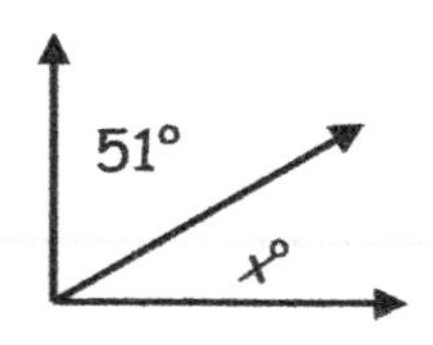

1.	2.	3.	4.
5.	6.	7.	8.
9.	10.	11.	12.
13.	14.	15.	16.
17.	18.	19.	20.

Lesson #36

1. $3{,}625 + 9{,}884 = ?$

2. Solve for y. $y - 28 = -50$

3. Solve for x. $3x + 2(x + 4) = -22$

4. $\begin{pmatrix} 6 & 9 \\ 7 & -4 \end{pmatrix} - \begin{pmatrix} 4 & 7 \\ -6 & -1 \end{pmatrix} = ?$

5. $\dfrac{-246}{6} = ?$

6. How many centuries are 600 years?

7. What is the P(H, H, T, T, H) on 5 flips of a coin?

8. 60% of what number is 42?

9. $29 \times 42 = ?$

10. $10\frac{2}{7} - 8\frac{6}{7} = ?$

11. Write $\frac{1}{8}$ as a decimal.

12. $-80 + (-41) = ?$

13. $9.2 - 7.385 = ?$

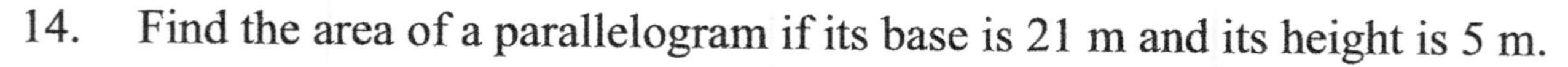

14. Find the area of a parallelogram if its base is 21 m and its height is 5 m.

15. $31 - (-18) = ?$

16. What is the value of n? $5n + 3n - 6 = 42$

17. On a local lake, the ratio of ducks to geese is 10 to 6. If there are 96 geese on the lake, how many ducks are there?

18. $5[3 + 2(4)] - 45 = ?$

19. Tell whether -28 is rational or irrational.

20. Write an algebraic expression that means *a number divided by 9, decreased by 3.*

1.	2.	3.	4.
5.	6.	7.	8.
9.	10.	11.	12.
13.	14.	15.	16.
17.	18.	19.	20.

Lesson #37

1. $-16(-4) = ?$

2. Find the value of x. $7x - 3 = 25$

3. A triangle with three congruent sides is a(n) _____________ triangle.

4. $-55 + (-34) = ?$

5. $\begin{pmatrix} -8 & 6 \\ 4 & -2 \end{pmatrix} + \begin{pmatrix} -6 & -4 \\ 5 & -1 \end{pmatrix} = ?$

6. What is the value of x? $5x = 125$

7. $0.7 - 0.246 = ?$

8. $\dfrac{-275}{-5} = ?$

9. Solve for x. $\dfrac{x}{9} = 15$

10. Find the GCF of $10xy^2$ and $12x^2yz^3$.

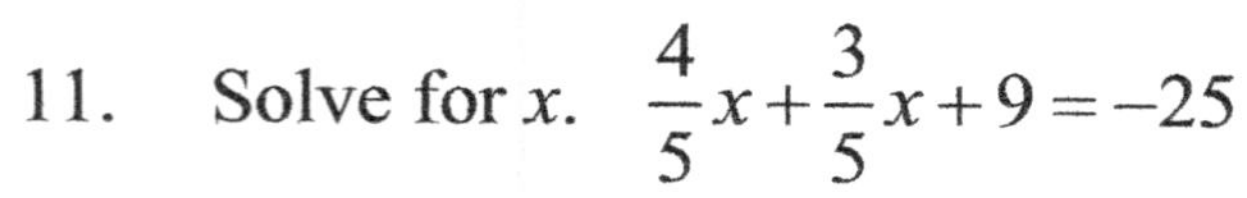

11. Solve for x. $\dfrac{4}{5}x + \dfrac{3}{5}x + 9 = -25$

12. $28\dfrac{1}{6} + 13\dfrac{2}{5} = ?$

13. Find the circumference of a circle whose diameter is 23 inches.

14. Find the value of x. $\dfrac{x}{4} + 3 = 13$

15. Rule for Addition of Integers: When the signs are the same, _________ and keep the sign. When the signs are different, ___________ and keep the sign of the larger number.

16. Solve to find the value of x. $\dfrac{1}{7}x + 8 = 20$

17. Evaluate $5x + 3y$ when $x = 5$ and $y = 4$.

18. $47 - (-21) = ?$

19. What value of x makes the equation true? $4x - 10 = 6x + 12$

20. Solve for n. $9(n + 7) = -81$

1.	2.	3.	4.
5.	6.	7.	8.
9.	10.	11.	12.
13.	14.	15.	16.
17.	18.	19.	20.

Lesson #38

1. $-7(-3)(2) = ?$

2. $17 - 13\frac{4}{7} = ?$

3. Solve for y. $8y - 2y + 3 = 9$

4. $192 + (-75) = ?$

5. Find the missing numbers in the function table.

$y = \frac{1}{2}x$

x	y
6	?
0	?
?	4

6. Determine the value of y. $\frac{3}{4}y = 9$

7. $-|-21| = ?$

8. The area of a square is 144 ft^2. What is the length of each side?

9. Solve for a. $a + 9 = 23$

10. 25% of 80 is what number?

11. Write 0.65 as a reduced fraction and as a percent.

12. What is the value of n? $3n + n - 8 = 32$

13. $-105 - (-36) = ?$

14. Solve for x. $5x + 6 = -24$

15. The average length of Marta's eight spelling words was 9 letters. The first seven words had varying lengths: 11 letters, 9 letters, 7 letters, 10 letters, 9 letters, 6 letters, and 8 letters. What was the length of the eighth word?

16. $42.7 + 8.623 = ?$

17. What is the percent of decrease from 24 feet to 12 feet?

18. Find the missing measurement, x.

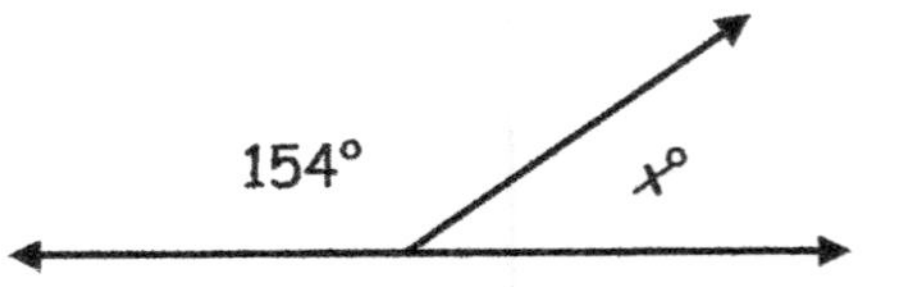

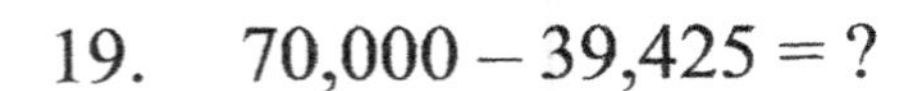

19. $70{,}000 - 39{,}425 = ?$

20. How many feet are in 6 miles?

1.	2.	3.	4.
5.	6.	7.	8.
9.	10.	11.	12.
13.	14.	15.	16.
17.	18.	19.	20.

Lesson #39

1. Solve for b. $\frac{b}{3} - 20 = 20$

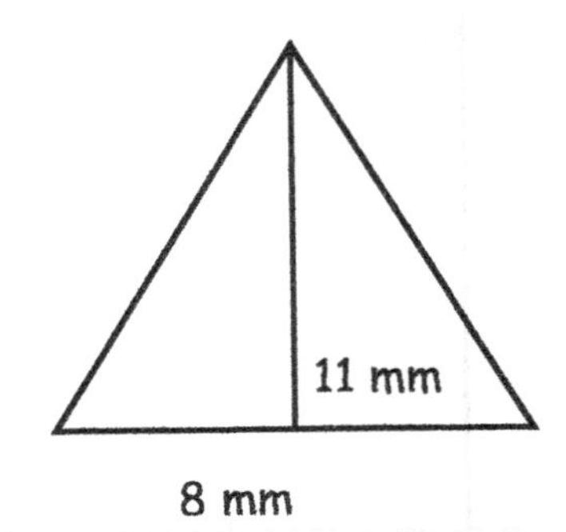

2. $0.005 \times 0.04 = ?$

3. Find the area of the triangle.

4. Find the value of x. $\frac{1}{4}x + 9 = 22$

5. $45 + (-21) = ?$

6. What is the value of a? $\frac{2}{5}a = -4$

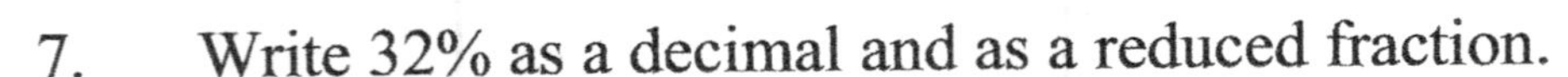

7. Write 32% as a decimal and as a reduced fraction.

8. Solve for n. $5n = -20$

9. How many years are in 6 centuries?

10. $-66 - (-17) = ?$

11. Find the value of a. $7a - 3 = 18$

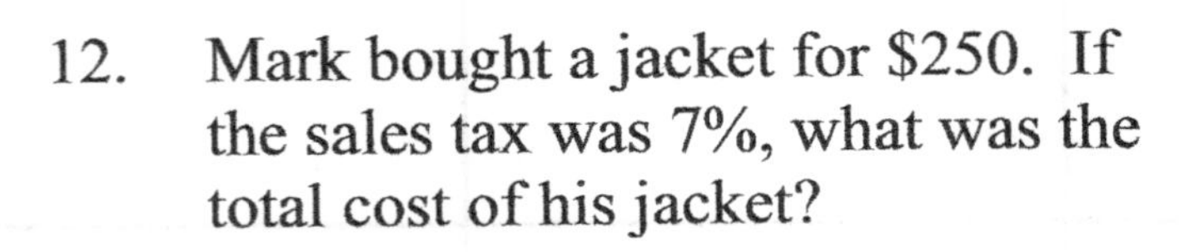

12. Mark bought a jacket for \$250. If the sales tax was 7%, what was the total cost of his jacket?

13. Solve for x. $\frac{x}{12} = 15$

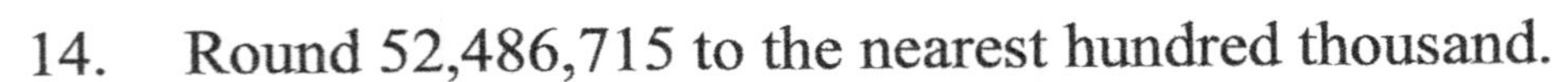

14. Round 52,486,715 to the nearest hundred thousand.

15. Solve to find the value of b. $2b + 5(b + 1) = -9$

16. $12\frac{1}{9} - 9\frac{7}{9} = ?$

17. If $a = 2$, $b = 4$ and $c = 5$, find the value of $\frac{abc}{2}$.

18. $2[6 + 2(4 + 7) - 8] = ?$

19. $-5(-9) = ?$

20. Distribute and simplify. $3(5x + 4y + 9) + 7(6y - 3)$

1.	2.	3.	4.
5.	6.	7.	8.
9.	10.	11.	12.
13.	14.	15.	16.
17.	18.	19.	20.

Lesson #40

1. Write 0.42 as a reduced fraction and as a percent.
2. $36{,}499 + 48{,}225 = ?$
3. $-46 + (-18) = ?$
4. Find the value of y. $11y + 9 = 130$
5. $-9(7) = ?$
6. A tree that measured 19 inches last spring grew to 25 inches this spring. The tree's growth represents what percent of increase?

7. $21 \div 3 \cdot 5 - 10 + 6 \div 2 = ?$
8. How many quarts are in 8 gallons?
9. What is the value of x? $\frac{x}{5} = 21$
10. Solve for m. $\frac{2}{3}m = 200$
11. $41 - (-10) = ?$
12. If Julio scored an average of 18 points per game in his first 5 basketball games, how many points did he score altogether in those games?
13. Solve to find the value of x. $5x - 9 = 16$
14. $\begin{pmatrix} 14 & -3 \\ 10 & 5 \end{pmatrix} - \begin{pmatrix} 8 & -7 \\ 3 & -8 \end{pmatrix} = ?$
15. Find the value of w. $2(w-7) = 90$
16. 90% of what number is 72?
17. What is the value of b? $b - 7 = -21$
18. Find the perimeter of a rectangle if it is 17 cm long and 4 cm wide.
19. Solve for a. $\frac{6}{7}a - \frac{5}{7}a + 3 = -12$
20. Find $\frac{4}{5}$ of 35.

1.	2.	3.	4.
5.	6.	7.	8.
9.	10.	11.	12.
13.	14.	15.	16.
17.	18.	19.	20.

Lesson #41

1. Solve for h. $10h-4=-94$
2. $4 \cdot 6 - 4 + 10 \div 2 + 2^2 = ?$
3. $52{,}310 - 28{,}477 = ?$
4. Solve for n. $6n=-90$
5. Find the value of x. $\frac{3}{4}x=6$
6. $-43 + 15 = ?$
7. $\frac{-55}{5} = ?$
8. Find the value of x. $6x-7=14x+9$

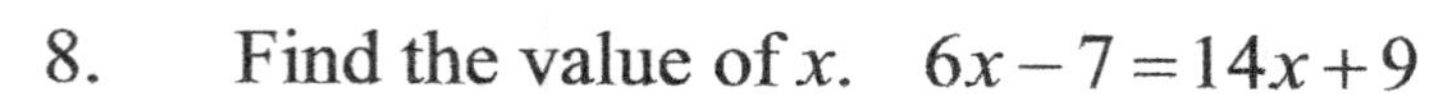

9. What is the value of w? $3w+2-w=-4$
10. $-54 - (-12) = ?$
11. Simplify. $7(3x+5y-2)$
12. Solve for y. $y-19=-52$
13. Write $\frac{4}{25}$ as a decimal and as a percent.

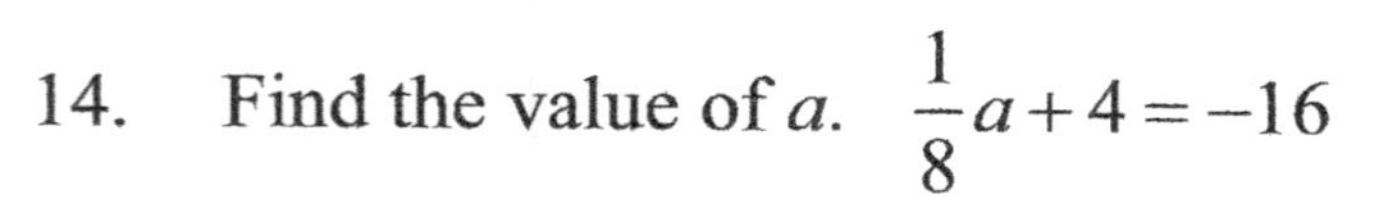

14. Find the value of a. $\frac{1}{8}a+4=-16$
15. $0.3 \div 1.5 = ?$

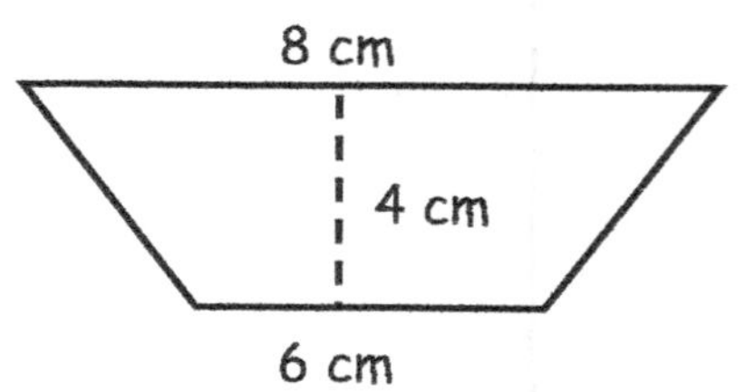

16. Calculate the area of the trapezoid.

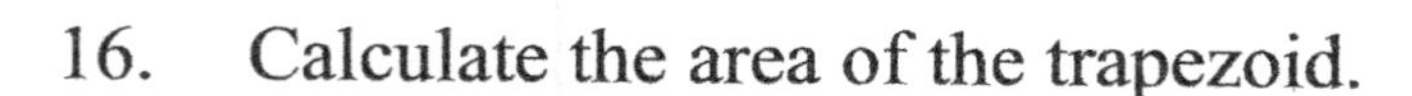

17. In the first year a puppy grew from 8 inches long to 14 inches long. What was the percent of increase in the puppy's length?
18. Find $\frac{3}{7}$ of 28.
19. $8 - 4\frac{2}{3} = ?$
20. Identify the missing numbers in the function table.

$y = 2x - 1$

x	y
5	?
3	?
1	?

1.	2.	3.	4.
5.	6.	7.	8.
9.	10.	11.	12.
13.	14.	15.	16.
17.	18.	19.	20.

Lesson #42

1. Solve to find the value of x. $3x = 168$

2. What is the value of y? $3y + 8 = -1$

3. Find the area of a parallelogram if the base is 15 meters and the height is 8 meters long.

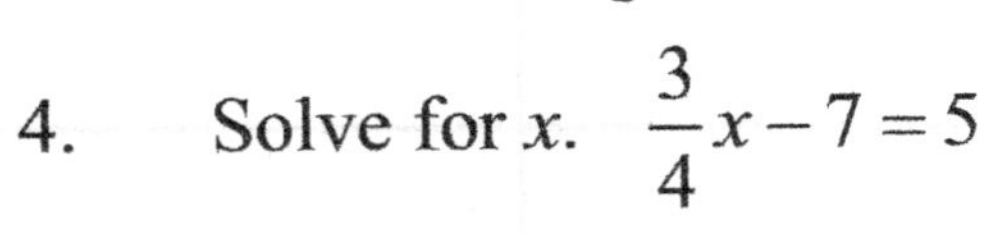

4. Solve for x. $\frac{3}{4}x - 7 = 5$

5. $3.6 \times 0.12 = ?$

6. $68 + (-42) = ?$

7. $\frac{4}{7} + \frac{2}{3} = ?$

8. $-8(5) = ?$

9. Find the value of x. $\frac{5}{7} = \frac{x}{98}$

10. At Paul's favorite Chinese restaurant, the main course costs \$9.50. He pays \$1.25 for his beverage and \$2.50 for soup, and he leaves a tip that is 20% of the price of his meal. How much does Paul leave for a tip?

11. $8\frac{2}{5} - 6\frac{4}{5} = ?$

12. Solve for x. $7x + 3(x + 5) = 25$

13. $-102 - (-47) = ?$

14. Write 30% as a decimal and as a fraction.

15. How many sides are in a decagon?

16. Solve to find the value of x. $\frac{x}{6} = 12$

17. $7{,}398 + 4{,}765 = ?$

18. The Morgan family put an addition on their house. The house was originally 32 feet long. What was the percent of increase in the length of their house if it is now 43 feet long? Round to the whole number.

19. What value of x makes the equation true? $5x - 3 = 2x + 12$

20. What percent of 90 is 27?

1.	2.	3.	4.
5.	6.	7.	8.
9.	10.	11.	12.
13.	14.	15.	16.
17.	18.	19.	20.

Lesson #43

1. Tell whether $\sqrt{5}$ is rational or irrational.
2. $41 + (-23) = ?$
3. Write $\frac{1}{4}$ as a decimal and as a percent.
4. Solve for x. $3x - 12 = 2x + 6$
5. $-9(-13) = ?$
6. Find the value of a. $-3a = 39$
7. $6[4 + 2^2 + 10 \div 2] = ?$
8. Solve for x. $\frac{x}{9} = 14$
9. $60{,}000 - 33{,}918 = ?$
10. What is the value of b? $b + 7 = 22$
11. Find the value of x. $3x + 9 = -9$
12. What is the P(6, 5) on 2 rolls of a die?
13. $-80 - (25) = ?$
14. $6\frac{1}{8} - 2\frac{5}{8} = ?$
15. $26.4 + 19.375 = ?$
16. Find the value of $\frac{5a}{b} + 2a$ when $a = 4$ and $b = 10$.
17. $96 \times 34 = ?$
18. $4\frac{3}{7} + 3\frac{1}{2} = ?$
19. The ratio of pies to cakes in the county baking contest is 3 to 5. If there are 75 cakes entered in the contest, how many pies are also entered?
20. What are the missing numbers in the function table?

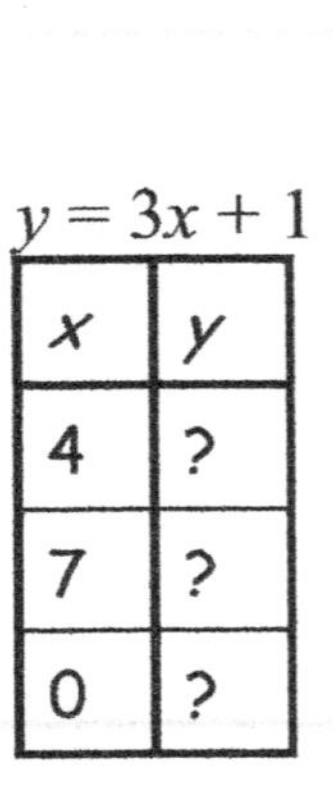

$y = 3x + 1$

x	y
4	?
7	?
0	?

1.	2.	3.	4.
5.	6.	7.	8.
9.	10.	11.	12.
13.	14.	15.	16.
17.	18.	19.	20.

Lesson #44

1. $6(-10) = ?$

2. $61 + (-10) = ?$

3. Solve for y. $5y = -45$

4. Simplify. $\dfrac{6x^2yz^4}{12xy^2z^3}$

5. What is the value of y? $y + 27 = 62$

6. Find the value of m. $4m + 2 = -18$

7. $121 - (-80) = ?$

8. What value of m makes the equation true? $6 + 3m = -m - 6$

9. $549{,}284 + 698{,}776 = ?$

10. $3^2 + 2[8 + 2(4 + 6) - 5] = ?$

11. Solve for x. $\frac{7}{9}x - \frac{6}{9}x + 14 = 25$

12. Find the value of d. $d - 4 = -9$

13. When $a = 2$ and $b = 4$, what is the value of $3ab - 2b$?

14. A triangle with two congruent sides is a(n) _________ triangle.

15. $-7^3 = ?$

16. Find the median and mode of the number of books read by the students at Lincoln Elementary.

Books Read by Students at Lincoln Elementary

Stem	Leaf
0	1 2 4
1	0 7
2	1 5 5
3	0

17. $6\frac{2}{5} + 3\frac{3}{7} = ?$

18. Write $\frac{11}{20}$ as a decimal and as a percent.

19. Simplify. $9(2a + 4b + 8) + 2(5a - 6)$

20. The Subtraction Rule for Integers: Change the sign of the ___________ number and _______________.

1.	2.	3.	4.
5.	6.	7.	8.
9.	10.	11.	12.
13.	14.	15.	16.
17.	18.	19.	20.

Lesson #45

1. Write an algebraic phrase for *five times a number increased by 8.*
2. –66 – (–38) = ?
3. What is the value of x? $\frac{x}{15} = 13$
4. Solve for a. $5a + 7 = 22$
5. Carmen wants to tie a piece of ribbon onto balloons for a party. She has a ribbon that is $22\frac{1}{2}$ inches long which she'll cut into three equal pieces. How long will each piece be?
6. Combine like terms. $3b + 2 + 5b$
7. Write 0.32 as a percent and as a reduced fraction.
8. The temperature on Sunday was 75°F. On Monday it was 90°F. What was the percent of change in the temperature from Sunday to Monday?
9. Find the value of x. $2x + 3x - 4 = 11$
10. 0.5 – 0.3342 = ?
11. Solve for x. $3x + x = 16$
12. Simplify. $12(s + 2t + 3w)$
13. What is the value of y? $3(5y - 8) = 7y$
14. Solve for t. $2t = -108$
15. 93 + (–29) = ?
16. Find the value of a. $a - 7 = -16$
17. 3[16 – (3 + 7) ÷ 5] = ?
18. $7\frac{2}{5} + 8\frac{1}{3} = ?$
19. Evaluate $(5 + b)^2 + a$ when $a = 2$ and $b = 7$.
20. $\begin{pmatrix} 8 & -4 & 2 \\ -6 & 5 & 9 \end{pmatrix} - \begin{pmatrix} 3 & -6 & 0 \\ -4 & 9 & 6 \end{pmatrix} = ?$

1.	2.	3.	4.
5.	6.	7.	8.
9.	10.	11.	12.
13.	14.	15.	16.
17.	18.	19.	20.

Lesson #46

1. Solve to find the value of m. $m+39=71$

2. Find the value of x. $2x-8=-20$

3. Write $\frac{2}{25}$ as a decimal and as a percent.

4. What is the value of x? $\frac{x}{2}+6=10$

5. Solve for x. $3x-5=x+3$

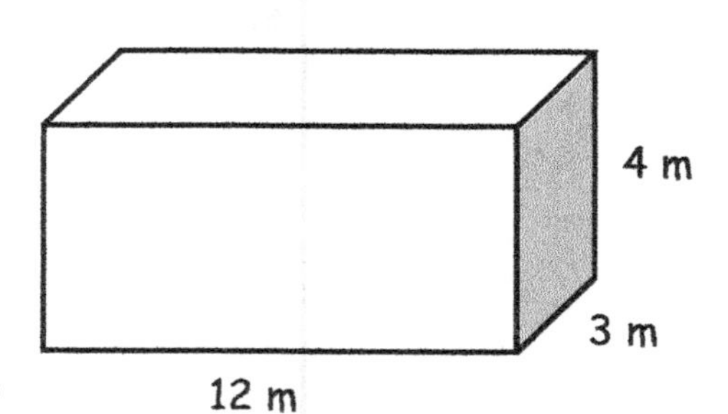

6. $8+14-2+4\cdot 2^3=?$

7. What is the value of t? $5t=65$

8. Find the volume of the rectangular prism.

9. Write 7.006 using words.

10. Simplify. $3a+4b+8+6a+9$

11. The rabbit to fox ratio in the state park is 9 to 7. If there are 91 foxes in the park, how many rabbits are there?

12. If $x=5$, $y=8$ and $z=9$, what is the value of $y(z-x)+x$?

13. Factor. $30a^4bc^2$

14. $\frac{120}{x}=40$ (Hint: Use a proportion.)

15. $47+(-13)=?$

16. $12\frac{2}{7}-9\frac{6}{7}=?$

17. What is the value of x? $x-12.42=9$

18. Find the perimeter of a regular pentagon whose sides measure 12 cm.

19. Solve for x. $\frac{x}{3}=36$

20. 60% of what number is 900?

1.	2.	3.	4.
5.	6.	7.	8.
9.	10.	11.	12.
13.	14.	15.	16.
17.	18.	19.	20.

Lesson #47

1. Determine the value of $(8-x)\div x$ when $x=4$.

2. Solve the equation for x. $6x=-102$

3. $37-(-13)=?$

4. Find the missing numbers in this function table.

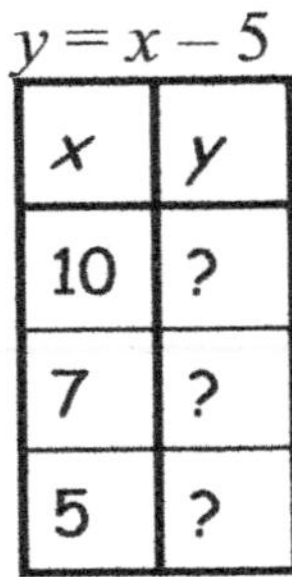

$y=x-5$

x	y
10	?
7	?
5	?

5. $13{,}467{,}871+47{,}866{,}938=?$

6. Simplify. $6(3m+4n-8)$

7. $21-1\cdot 2+5-3^2=?$

8. Solve for b. $b+7=19$

9. Write an algebraic expression that means *ten less than a number.*

10. Find the value of 9^5.

11. $-40+(-11)=?$

12. Find the value of d. $\dfrac{d}{12}=4$

13. $-|14|=?$

14. Sixteen teams are playing in a volleyball tournament. During the first round, each team will play one game against another. There is one gym, so only one game can be played at a time. If the first round must be completed in 6 hours, what is the maximum length of time that can be allowed for each game?

15. What is the value of x? $\dfrac{x}{4}=\dfrac{32}{64}$

16. Which is greater, 76% or $\dfrac{16}{25}$?

17. Simplify. $3(2x+3)+4x$

18. Put these integers in increasing order.

    –16, 41, –1, 12, 0

19. Solve for x. $\dfrac{x}{2}+5=10$

20. Find the width of a rectangle if its area is 36 ft^2 and its length is 12 ft.

1.	2.	3.	4.
5.	6.	7.	8.
9.	10.	11.	12.
13.	14.	15.	16.
17.	18.	19.	20.

Lesson #48

1. $60{,}000 - 41{,}219 = ?$

2. Write the formula for finding the surface area of a rectangular prism.

3. $7[8 + 2 \cdot 3 - 5] = ?$

4. Solve for m. $\frac{m}{2} + 2 = -1$

5. Find the LCM of $16a^2bc^3$ and $18abc$.

6. What is the value of y? $6y - 3y + 2 = -16$

7. $2(x-9) = 3(x-6)$ (Hint: Distribute first, then solve for x.)

8. $\frac{4}{5} + \frac{2}{3} = ?$

9. Find the value of n. $10(2n+10) = 120n$

10. Simplify. $2x + 4y + 3z + 16z$

11. Find the y-values in the function $y = 7x + 1$ when the domain is $\{2, 0, -1\}$.

12. Solve for x. $3(x+7) = 27$

13. Find the value of y. $\frac{y}{6} = 345$

14. $15 - 8\frac{5}{7} = ?$

15. Write *a number divided by 5 is 12* using algebraic symbols.

16. Find the value of x. $-5x = 625$

17. If $a = 2$ and $b = 7$, find $b(9-a)$.

18. Solve for y. $y - 219 = 356$

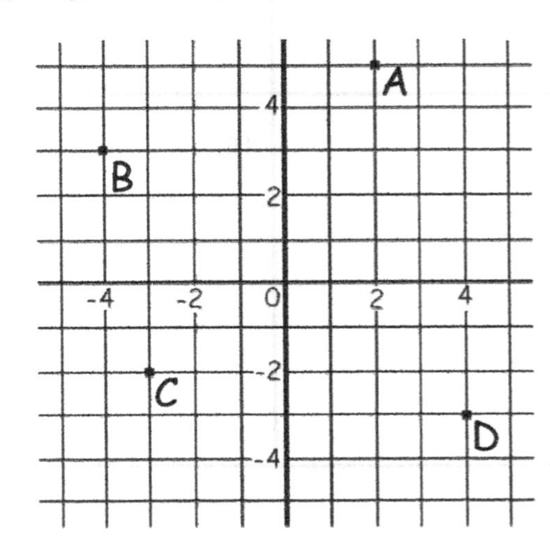

Give the coordinates of each point.

19. A  B______

20. C ______ D 

1.	2.	3.	4.
5.	6.	7.	8.
9.	10.	11.	12.
13.	14.	15.	16.
17.	18.	19.	20.

Lesson #49

1. What is the value of x? $\frac{x}{7} = 21$
2. $\frac{-75}{5} = ?$
3. A scout troop put 12 markers along a hiking trail. Each marker weighed 24 ounces. When the scouts began the hike, they put the markers in a backpack. If the empty backpack weighed 4 pounds, how much did the backpack weigh with all of the markers in it?

4. $101 + (-69) = ?$
5. Solve for c. $c + 316 = 907$
6. $16 - 2[3 + 4(2 + 1) - 5] = ?$
7. $\frac{15}{16} \times \frac{4}{5} = ?$
8. Simplify. $6(4a + 3b) + 2(5a - 6)$
9. $133 - (57) = ?$
10. Find the value of y. $4(7 + y) = 16 - 2y$
11. Find the GCF of $3x^2y^2$ and $15x^2y$.
12. Express the phrase *three more than twice a number* using algebraic symbols.
13. $7(-4)(-2) = ?$
14. Evaluate $3a^2 \cdot b$ when $a = 2$ and $b = 7$.
15. Find the circumference of a circle with a diameter of 25 mm.
16. Solve for y. $\frac{26}{y} = 2$
17. Find the value of a. $8a - 2a + 12 = 42$
18. $3,652 \times 8 = ?$
19. What is the value of x? $-11x = 77$
20. Find the missing values in the function chart.

$y = 2x + 3$

x	y
1	?
0	?
-2	?

1.	2.	3.	4.
5.	6.	7.	8.
9.	10.	11.	12.
13.	14.	15.	16.
17.	18.	19.	20.

Lesson #50

1. $[1 + 3(8 + 6)] - 4^2 = ?$

2. Put these decimals in increasing order. 0.7, 0.563, 0.24, 0.78

3. When $a = 2$ and $b = 3$, find the value of $9a - 4b + 6$.

4. The measures of complementary angles add up to ________ degrees.

5. Simplify. $5(2a + 4b) + 3(4a - 2b)$

6. What kind of graph is shown to the right?

Miles Hiked in September

Stem	Leaf
1	3 5 7
2	0 2
3	1

7. $205 - (-86) = ?$

8. Factor. $12x^3y^2$

9. Find the area of the trapezoid.

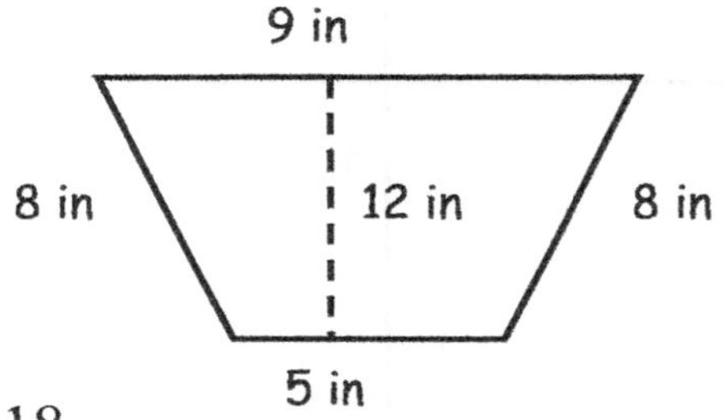

10. Solve for m. $6m - 2m - 4 = -60$

11. $119 + (-58) = ?$

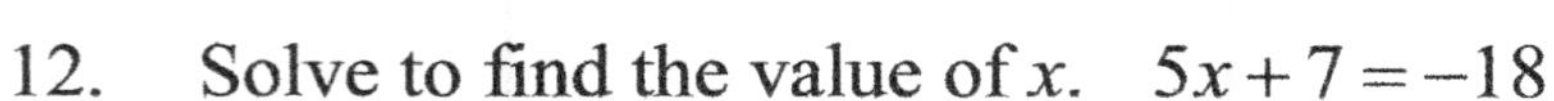

12. Solve to find the value of x. $5x + 7 = -18$

13. Find the value of p. $p + 91 = 162$

14. Round 47.623 to the nearest tenth.

15. Find the percent of change from 21 inches to 54 inches.

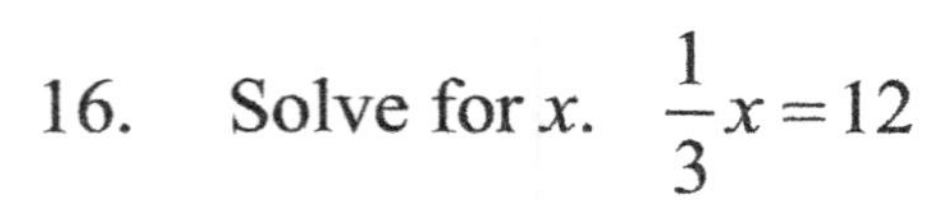

16. Solve for x. $\frac{1}{3}x = 12$

17. $-16 + (-14) + 12 = ?$

18. $0.15 \times 0.005 = ?$

19. In Maria's garden, the ratio of peppers to tomatoes is 5 to 7. If there are 91 tomatoes, how many peppers are in the garden?

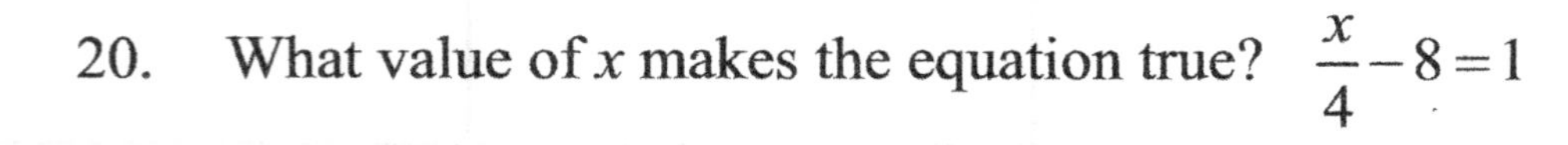

20. What value of x makes the equation true? $\frac{x}{4} - 8 = 1$

1.	2.	3.	4.
5.	6.	7.	8.
9.	10.	11.	12.
13.	14.	15.	16.
17.	18.	19.	20.

Lesson #51

1. $182 - (-76) = ?$

2. Find the value of x. $3x + 9 = -18$

3. Solve for x. $\frac{x}{5} + 9 = -18$

4. $9 - 7\frac{2}{7} = ?$

5. What value of x makes the equation true? $5x - 3 = 2x + 12$

6. Write 70% as a decimal and as a fraction.

7. $\frac{-2,436}{6} = ?$

8. Find the GCF of $2y^2z$ and $8yz^2$.

9. $32\frac{5}{6} + 18\frac{1}{4} = ?$

10. $100 + (-59) = ?$

11. 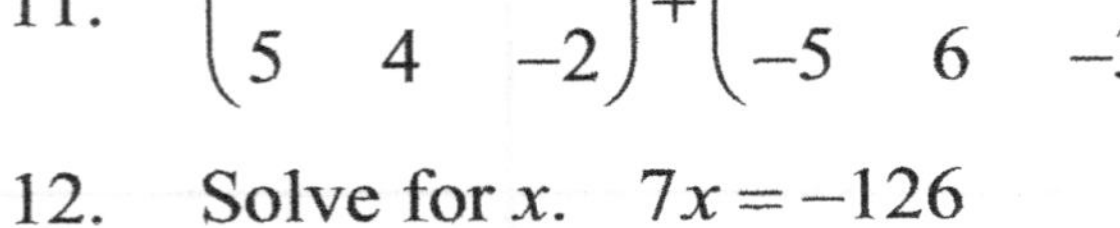

$$\begin{pmatrix} 7 & -3 & 0 \\ 5 & 4 & -2 \end{pmatrix} + \begin{pmatrix} 6 & -2 & 9 \\ -5 & 6 & -3 \end{pmatrix} = ?$$

12. Solve for x. $7x = -126$

13. The cost for parking in a downtown lot can be figured using the following formula: $cost = 3 + 2(time - 1)$. What is the cost of parking for 4 hours? (Hint: Cost is measured in dollars, and time is in hours.)

14. The measures of supplementary angles add up to __________ degrees.

15. $1\frac{1}{2} \times 2\frac{1}{3} = ?$

16. $9 + 2[6 + 2(3)] = ?$

17. Solve for x. $x - 17 = -41$

18. When $a = 2$, $b = 3$ and $c = 4$, determine the value of $a(b + c)$.

19. On the Fahrenheit temperature scale, water boils at _______________.

20. $\sqrt{784} - \sqrt{49} + 3^2 = ?$

1.	2.	3.	4.
5.	6.	7.	8.
9.	10.	11.	12.
13.	14.	15.	16.
17.	18.	19.	20.

Lesson #52

1. $-4(-5)(2) = ?$
2. $8 + 3[5 + 2(3) - 2] = ?$
3. $\frac{6}{7} \times \frac{14}{18} = ?$
4. Simplify. $12(2x+3y-5)+4(6y-4)$
5. Factor. $30a^3b^2c$
6. $-16 + (-14) + 10 = ?$
7. Solve to find the value of x. $-7x = 147$
8. Find the value of x. $\frac{1}{5}x = 15$

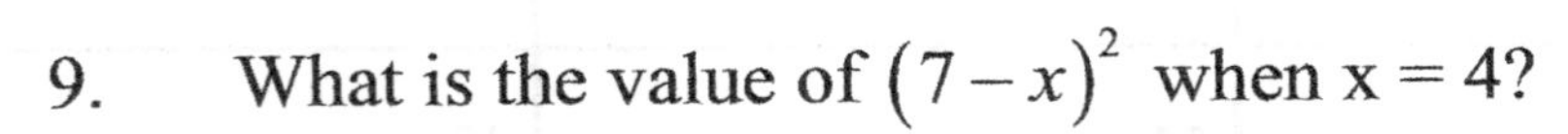

9. What is the value of $(7-x)^2$ when x = 4?

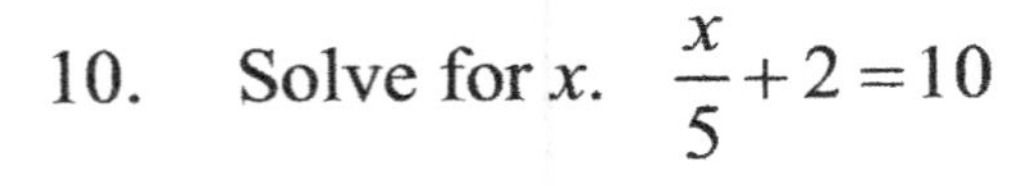

10. Solve for x. $\frac{x}{5} + 2 = 10$

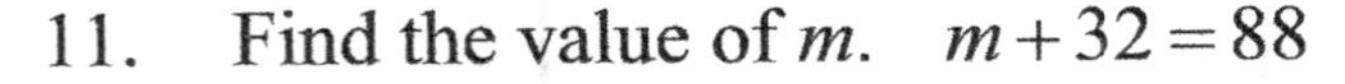

11. Find the value of m. $m + 32 = 88$
12. Find the GCF of $2x^2y^3$ and $4xy^8$.
13. Translate *four divided by a number* into an algebraic expression.
14. Solve for a. $5a - 3a + 12 = 8$
15. Find the value of x. $x - 14.4 = 7.02$
16. Solve for a. $\frac{a}{16} = 9$
17. What is the value of n? $4n - n = n + 9$
18. $\frac{5}{8} = \frac{?}{200}$
19. Find the volume of a rectangular prism whose length is 7 mm, width is 5 mm, and whose height is 3 mm.

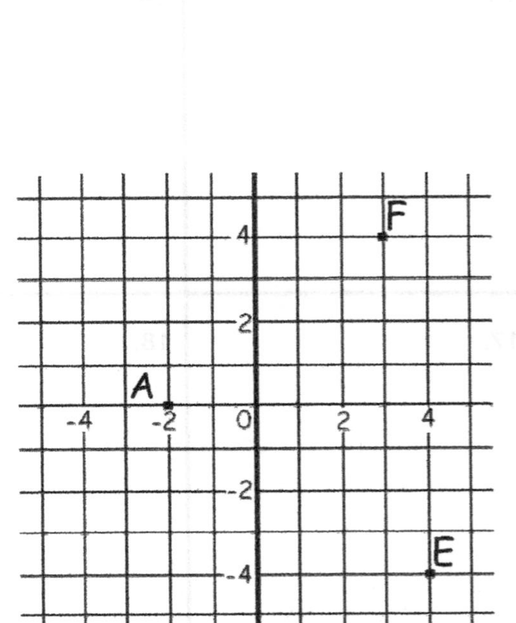

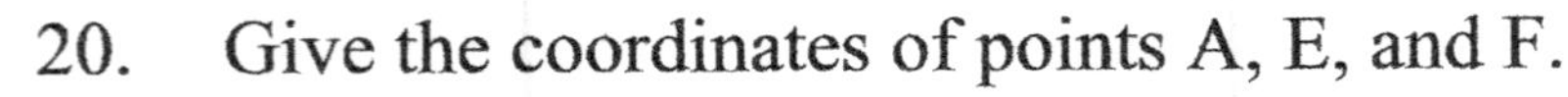

20. Give the coordinates of points A, E, and F.

1.	2.	3.	4.
5.	6.	7.	8.
9.	10.	11.	12.
13.	14.	15.	16.
17.	18.	19.	20.

Lesson #53

1. Determine the value of x. $2x+4=x-6$

2. Solve for a. $a-92=-126$

3. Find the surface area of this prism.

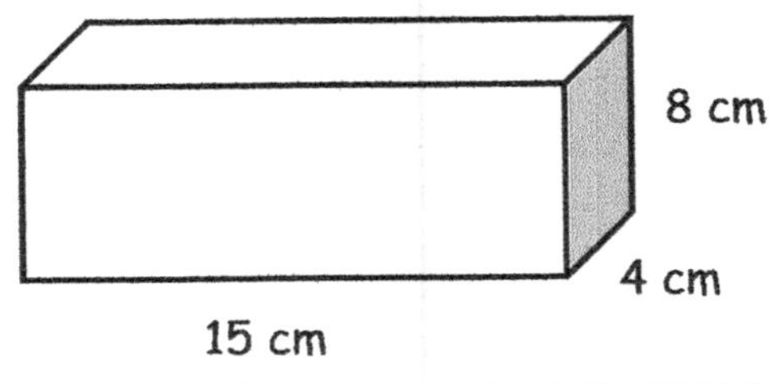

4. $82+(-34)=?$

5. Find the value of x. $\frac{4}{5}x-\frac{3}{5}x+9=12$

6. Find the GCF of $16a^3b^4c^2$ and $24ab^2c$.

7. $152-(-64)=?$

8. Solve for x. $|x-3|=8$

9. Evaluate $a(2b-c)$ when $a=2$, $b=3$ and $c=4$.

10. $9\frac{2}{7}+3\frac{2}{3}=?$

11. Solve for x. $\frac{1}{6}x-8=-25$

12. $72\div 9+8\div 4\cdot 2-\sqrt{64}=?$

13. Write 18% as a decimal and as a reduced fraction.

14. Find the value of y. $y+123=216$

15. $-4(-9)(2)=?$

16. What is the value of x? $\frac{x}{7}+8=14$

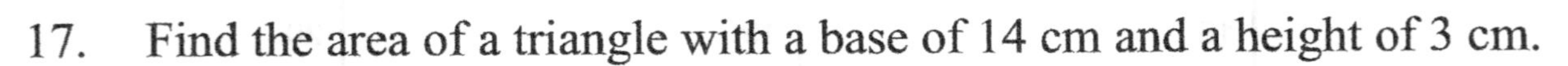

17. Find the area of a triangle with a base of 14 cm and a height of 3 cm.

18. What value of x makes the equation true? $3x+8=-16$

19. Solve for a. $16a=-112$

20. Carlos' weight went from 185 pounds to 141 pounds in 6 months. By what percent did Carlos' weight change? Round to the nearest whole number.

1.	2.	3.	4.
5.	6.	7.	8.
9.	10.	11.	12.
13.	14.	15.	16.
17.	18.	19.	20.

Lesson #54

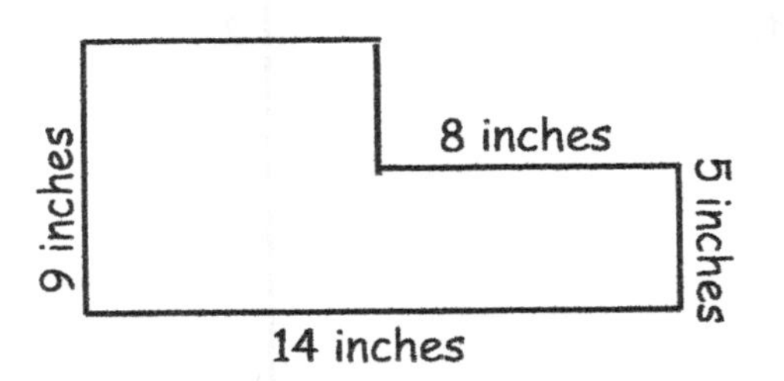

1. Find the perimeter of the complex shape.

2. Find the value of x. $6x+7=-23$

3. Solve for a. $\frac{1}{9}a-12=24$

4. $\sqrt{81}\div 3+15-6+2=?$

5. What is the value of y? $28+4y=16-2y$

6. Solve for n. $5(2n+3)=65$

7. When $x=3$, $y=2$ and $z=5$, evaluate $x(y+z)$.

8. $-129-(-88)=?$

9. Find the value of c. $c-8=39$

10. A square has a side that measures 13 inches. What is its area?

11. $6\div 1.5=?$

12. Solve for x. $\frac{x}{6}-8=22$

13. $\frac{8}{10}\div\frac{2}{5}=?$

14. $119+(-87)=?$

15. Tell whether $\frac{2}{5}$ is rational or irrational.

16. What is the value of a? $\frac{1}{5}a=25$

17. Find the LCM of $2x^3y$ and $3xy^5$.

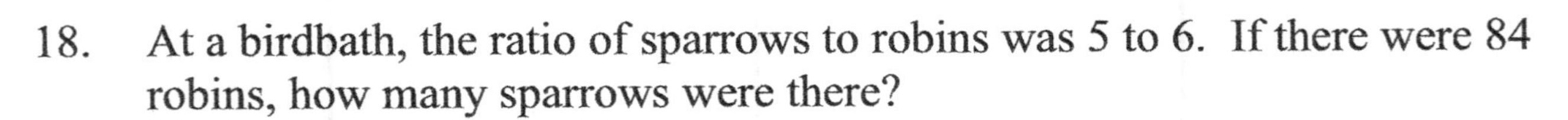

18. At a birdbath, the ratio of sparrows to robins was 5 to 6. If there were 84 robins, how many sparrows were there?

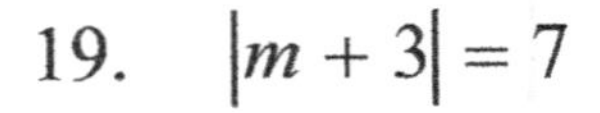

19. $|m+3|=7$

20. $\frac{4}{5}\times\frac{15}{20}=?$

1.	2.	3.	4.
5.	6.	7.	8.
9.	10.	11.	12.
13.	14.	15.	16.
17.	18.	19.	20.

Lesson #55

1. $-127 + (-65) = ?$

2. If Rachael bought five pounds of bananas for \$1.20, what was the price per pound? At this price, what will be the cost of 6 pounds of bananas?

3. Simplify. $7(2a+6b-9)$

4. When $x = 2$ and $y = 6$, what is the value of $(9x-y)^2$?

5. Solve for x. $5x+7=-18$

6. What is the value of c? $c-68=-135$

7. Find the LCM of $4a^6b^3$ and $8a^7b^5$.

8. $9[4+7-3]+\sqrt{25} = ?$

9. $34\frac{3}{7} + 19\frac{2}{3} = ?$

10. $7.2 \times 0.04 = ?$

11. A heptagon has _______ sides.

12. $8{,}000{,}000 - 6{,}476{,}293 = ?$

13. Find the value of x. $x+112=213$

14. Solve for x. $\frac{x}{7}=19$

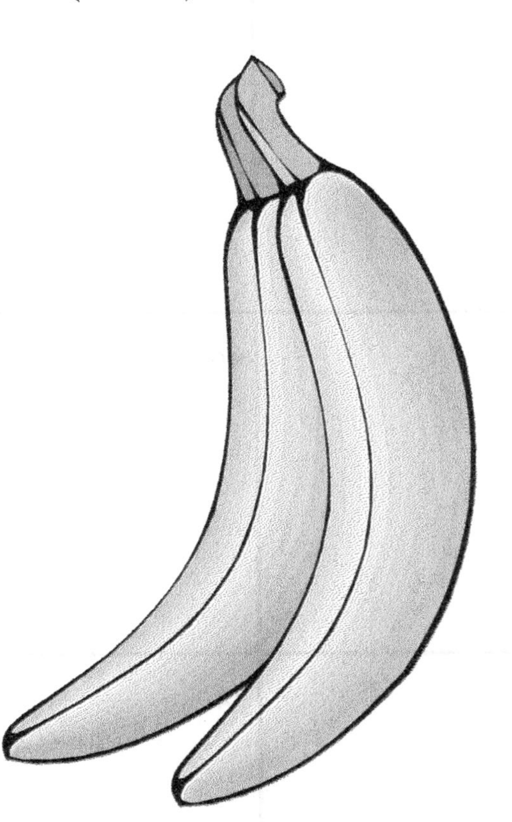

15. What is the value of x? $\frac{1}{8}x-10=21$

16. Write an expression that means *six times a number increased by 12.*

17. $96 + (-20) = ?$

18. Find the value of x. $\frac{1}{6}x=12$

19. Evaluate $a+2b-c$ when $a=2$, $b=3$ and $c=1$.

20. $\begin{pmatrix} 5 & 7 & 0 \\ 3 & -4 & -6 \end{pmatrix} - \begin{pmatrix} 2 & 0 & -1 \\ 4 & -5 & -3 \end{pmatrix} = ?$

1.	2.	3.	4.
5.	6.	7.	8.
9.	10.	11.	12.
13.	14.	15.	16.
17.	18.	19.	20.

Lesson #56

1. $8(2 + 3) - 5 + 10 \div 5 = ?$

2. Find the GCF of $10a^3b^2c^2$ and $15a^2bc^2$.

3. Find the value of h. $h - 79 = -100$

4. What is the value of a? $3a + 12 = -24$

5. Solve for x. $\frac{x}{9} + 9 = -9$

6. Factor. $56x^4y^2z$

7. Lisa bought a bottle of hand lotion that cost $3.55. If she paid with a ten-dollar bill, what is the fewest number of bills and coins she could have received back as change?

8. Find the value of x. $\frac{1}{4}x + 8 = -19$

9. A rectangular prism has a length of 13 mm, a width of 10 mm, and a height of 2 mm. Calculate the volume of the figure.

10. Solve for x. $9x + 12 = 15x$

11. Translate *the sum of a number and 17* into an algebraic expression.

12. Find $3(x + y) - 2x$ if $x = 4$ and $y = 5$.

13. $36 \div 9 - 6 \div 2 + 12 = ?$

14. Write 56% as a decimal and as a reduced fraction.

15. What is the average of 15, 18 and 21.

16. Find the value of d. $|2d + 7| = 11$

17. Solve for z. $-4z = 64$

18. $\frac{-90}{9} = ?$

19. Round 46,821 to the nearest thousand.

20. Solve to find the value of x. $2x + 3x - 4 = 11$

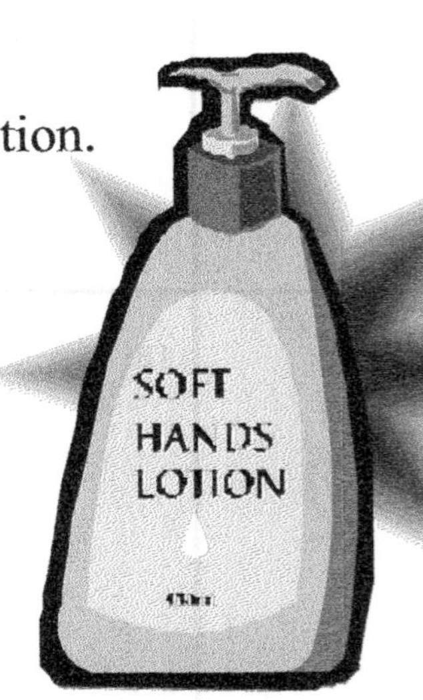

1.	2.	3.	4.
5.	6.	7.	8.
9.	10.	11.	12.
13.	14.	15.	16.
17.	18.	19.	20.

Lesson #57

1. Solve for x. $\frac{x}{12}=4$

2. *Four divided by a number is 12.* Rewrite the sentence as an equation.

3. What is the value of x? $x-4=19$

4. $3(5+2)-11+3^2=?$

5. Find the value of x. $\frac{24}{x}=3$

6. What is the value of y? $-4y=68$

7. $-41+(-19)+15=?$

8. Find the LCM of $7s^2t$ and $49st^2$.

9. Solve for m. $m-5.86=7.49$

10. $\frac{5}{13}$ ◯ $\frac{7}{12}$

11. Combine like terms. $5a+3b+7a+9+2b+8$

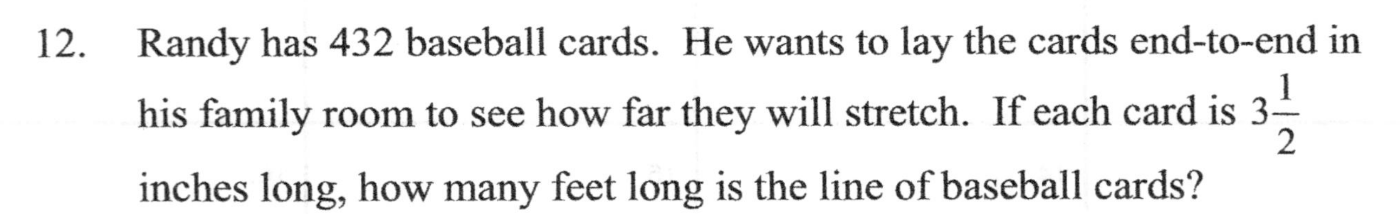

12. Randy has 432 baseball cards. He wants to lay the cards end-to-end in his family room to see how far they will stretch. If each card is $3\frac{1}{2}$ inches long, how many feet long is the line of baseball cards?

13. What is the perimeter of this regular hexagon?

9 ft.

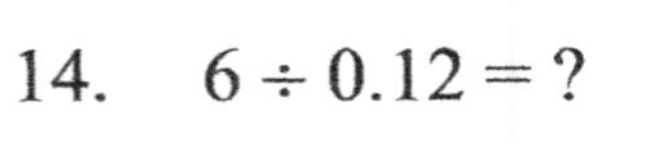

14. $6\div 0.12=?$

15. Solve for x. $\frac{1}{2}x=17$

16. Find the value of x. $7x+14=-14$

17. $\frac{6}{14}\div\frac{2}{7}=?$

18. Write $\frac{2}{50}$ as a decimal and as a percent.

19. 40% of what number is 60?

20. In the function table, what are the missing y-values?

$y=-2x$

x	y
3	?
0	?
-2	?

1.	2.	3.	4.
5.	6.	7.	8.
9.	10.	11.	12.
13.	14.	15.	16.
17.	18.	19.	20.

Lesson #58

1. Solve for a. $a+25=-100$
2. What is the value of y? $6y-3y+4=-11$
3. Find the GCF of $15a^3b^4c^2$ and $30a^2b^2c$.

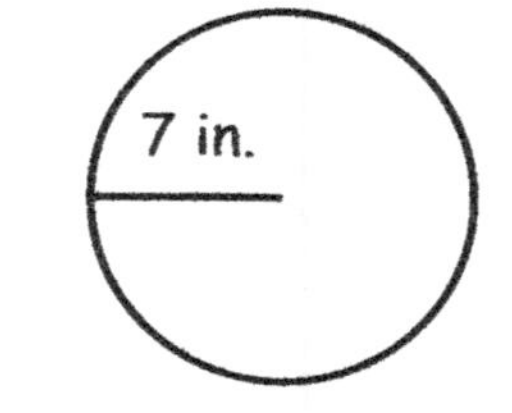

4. Calculate the area of the circle shown.
5. What value of x makes the equation true? $-5x+6=x+12$
6. Find the value of x. $\frac{1}{10}x-12=-42$
7. If $a=2$, $b=8$ and $c=3$, what is the value of $b-a\cdot c$?
8. $10{,}000-6{,}441=?$
9. Solve for x. $\frac{x}{5}+6=11$

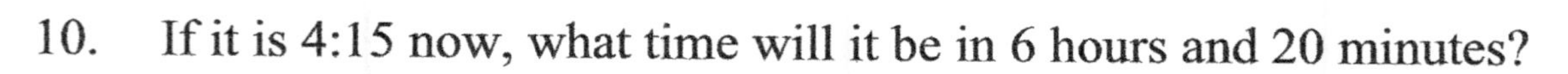

10. If it is 4:15 now, what time will it be in 6 hours and 20 minutes?
11. Find $\frac{4}{5}$ of 100.

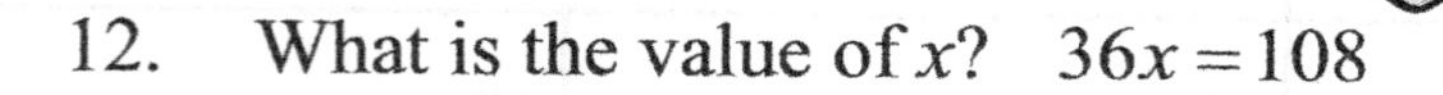

12. What is the value of x? $36x=108$
13. Find the value of m. $m-125=215$

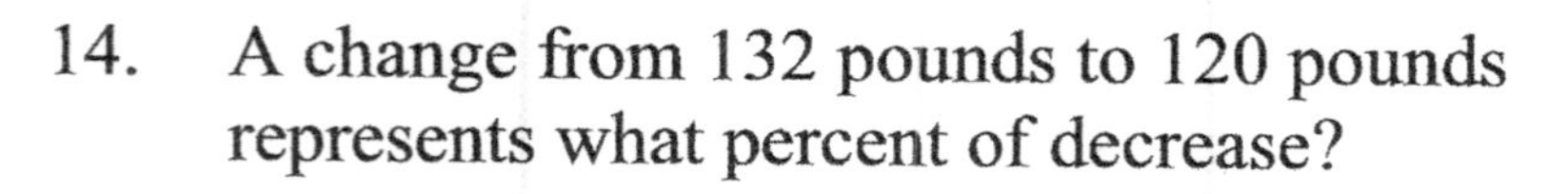

14. A change from 132 pounds to 120 pounds represents what percent of decrease?
15. $37+(-17)+24=?$

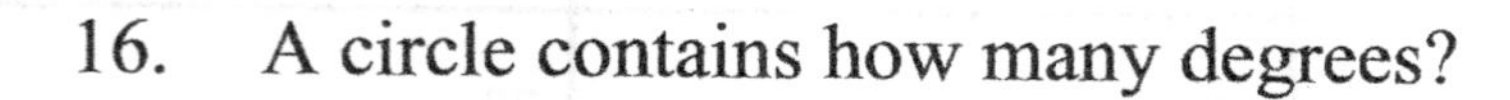

16. A circle contains how many degrees?
17. Combine like terms. $16x+4y+10+4x-2y$
18. What is the value of t? $|13-2t|=5$
19. Solve for x. $4x+3=19$
20. $-105-(-55)=?$

1.	2.	3.	4.
5.	6.	7.	8.
9.	10.	11.	12.
13.	14.	15.	16.
17.	18.	19.	20.

Lesson #59

1. Order these integers from greatest to least. $-2,\ 0,\ -19,\ -40$

2. $-214 + (-76) = ?$

3. Find the LCM of $6a^3b^4c^2$ and $15a^2b^3c$.

4. What is the value of x? $\frac{1}{4}x + 3 = 15$

5. Write 0.06 as a reduced fraction and as a percent.

6. Find the value of x. $\frac{x}{12} = 6$

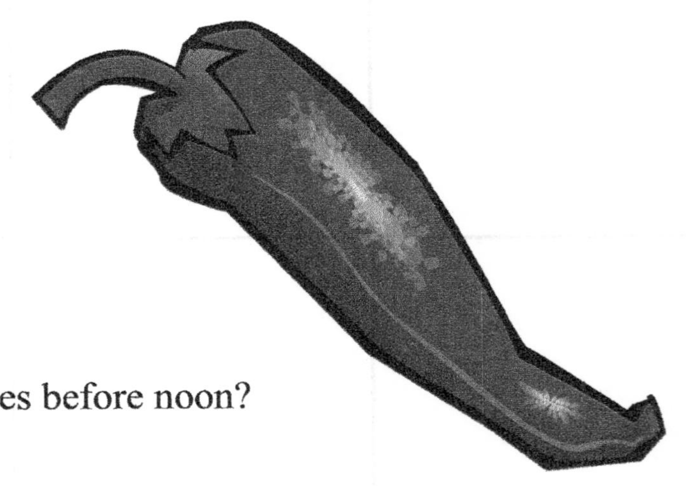

7. Solve for y. $-14y = 42$

8. $142 - (-96) = ?$

9. Solve for x. $8(3x + 2) = -56$

10. What will be the time 90 minutes before noon?

11. $16 - 2[3 + 2(4)] = ?$

12. What value of x makes the equation true? $10x + 17 = 4x - 1$

13. Find the median and the mode of 18, 25, 9, 18 and 66.

14. What is $a^2 + b^2$ when $a = 2$ and $b = 3$?

15. $-4(9) = ?$

16. Solve for a. $\frac{1}{2}a + 6 = 8$

17. 30 is what percent of 60?

18. According to the pie chart, what percent of the family's income is spent on food? What fraction of the income is spent on housing?

food
auto 12%
taxes 20%
insurance 10%
housing 25%
miscellaneous 20%

19. Solve for m. $6m + 5 = -1$

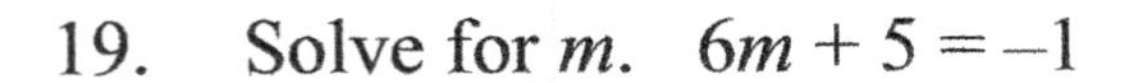

20. A rectangle with a length of 25 cm and a width of 5 cm has what area?

1.	2.	3.	4.
5.	6.	7.	8.
9.	10.	11.	12.
13.	14.	15.	16.
17.	18.	19.	20.

Lesson #60

1. $46 + (-21) = ?$

2. The distance across a circle through the center is the __________.

3. Find the value of x. $\frac{x}{9} = -2$

4. Solve for y. $4y + 10 = 6y$

5. $65{,}472{,}886 + 78{,}516{,}925 = ?$

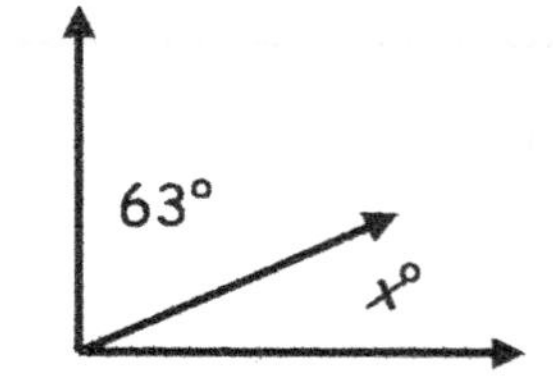

6. Find the missing measurement, x.

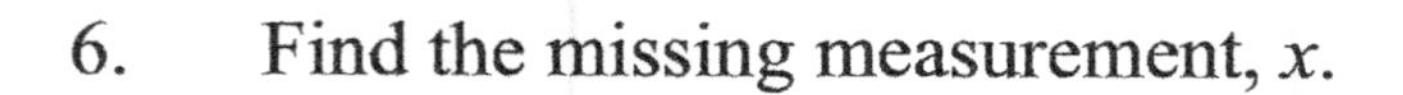

7. What is the value of x? $3x + 12 = -15$

8. What is the P(2, 1, 6, 3) on 4 rolls of a die?

9. $-162 - (-47) = ?$

10. Write 0.14 as a reduced fraction and as a percent.

11. Write an expression for *a number minus five.*

12. $(-9)(3)(-2) = ?$

13. Solve for b. $b + 91 = -156$

14. What is the range of the data in the stem-and-leaf plot?

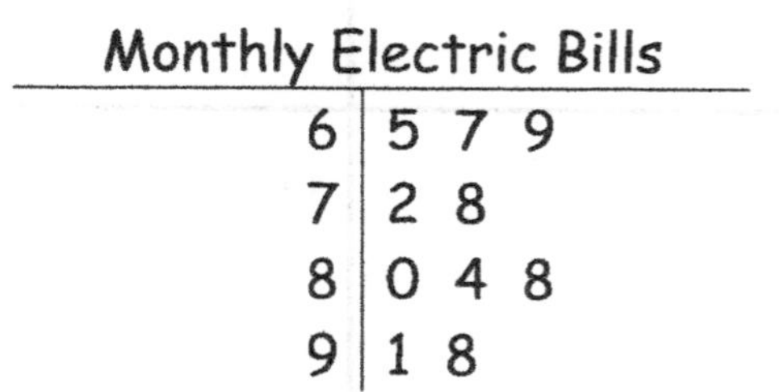

Monthly Electric Bills

Stem	Leaf
6	5 7 9
7	2 8
8	0 4 8
9	1 8

15. When $x = 2$ and $y = 5$, find $3x + y - x$.

16. $5^2 - 5 + 12 \div 3 + 2 = ?$

17. Solve for x. $\frac{x}{7} + 3 = 13$

18. Round 7.0364 to the nearest hundredth.

19. $19\frac{1}{10} - 15\frac{4}{5} = ?$

20. $52.8 \div 2.4 = ?$

1.	2.	3.	4.
5.	6.	7.	8.
9.	10.	11.	12.
13.	14.	15.	16.
17.	18.	19.	20.

Lesson #61

1. $-104-(-62)=?$

2. $24 \div 6 \cdot 9 - 6 + 15 \div 5 = ?$

3. What time was it 6 hours and 15 minutes ago if it is 5:20 p.m. now?

4. Simplify. $\dfrac{3x^2y}{9y}$

5. $9\frac{1}{8}+6\frac{2}{3}=?$

6. Use algebraic symbols to represent *five times a number plus two times another number.*

7. Determine the value of $3a(b+c)$ when $a=2$, $b=3$ and $c=5$.

8. $82{,}416+78{,}395=?$

9. Put the integers in increasing order. $-92,\ 16,\ 4,\ 0$

10. Solve for a. $a-63=-112$

11. What is the value of x? $\dfrac{x}{7}=22$

12. What value of x makes the equation true? $3x-5=x+3$

13. A bag contains 3 red marbles, 5 blue marbles, and 2 green marbles. If you pick two marbles from the bag, replacing the first one before choosing the second one, what is the P(red and green)? P(blue and red)?

14. Find the GCF of $12a^2b^3$ and $18a^3b^2c$.

15. Solve for x. $4x+11=35$

16. Write $\dfrac{1}{8}$ as a decimal.

17. $98+(-27)=?$

18. Find the value of a. $5a+11a-6=74$

19. What is the value of x? $\dfrac{1}{8}x+9=13$

20. Solve for m. $5(m+5)=-50$

1.	2.	3.	4.
5.	6.	7.	8.
9.	10.	11.	12.
13.	14.	15.	16.
17.	18.	19.	20.

Lesson #62

1. Combine like terms. $4a+9+3b+12+6a$

2. Solve to find the value of x. $7x-5x+9=21$

3. What is the value of a? $\frac{1}{7}a+12=32$

4. $19.2 + 7.864 = ?$

5. Solve for x. $\frac{1}{5}x=15$

6. Find the value of a. $a+15=39$

7. $\sqrt{144}-\sqrt{81}+3^2=?$

8. What value of x makes the equation true? $\frac{x}{8}-7=25$

9. $9\frac{2}{5}-6\frac{4}{5}=?$

10. Find the LCM of $7s^2t$ and $12st^2$.

11. A watch was marked down from \$120 to \$60. Find the percent by which the cost of the watch went down.

12. $7 + 2[3 + 4(2)] = ?$

13. $0.008 \times 0.007 = ?$

14. Solve for n. $2n+n=n+8$

15. Evaluate $4a+2b-b$ when $a=2$ and $b=3$.

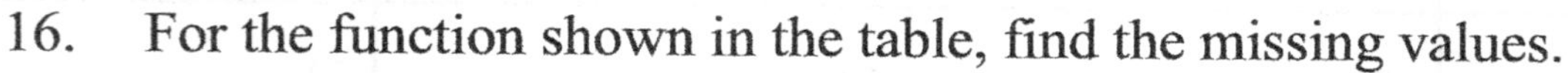

16. For the function shown in the table, find the missing values.

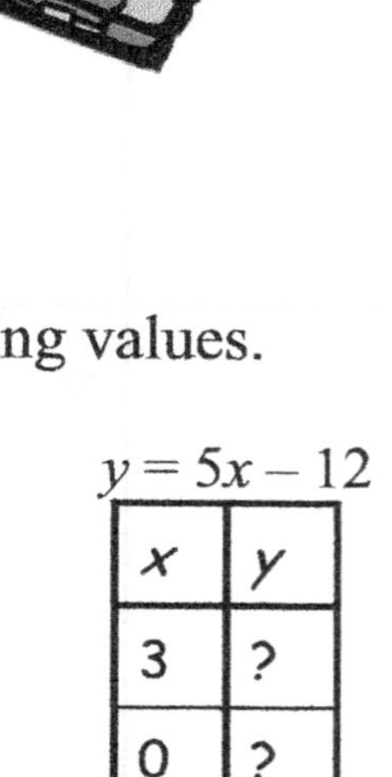

$y = 5x - 12$

x	y
3	?
0	?
-1	?

17. Simplify. $\frac{2ab}{8b^2}$

18. Solve for n. $3(n+4)+1=28$

19. Round 42,816,375 to the nearest ten thousand.

20. What is the value of x? $3x-4=11$

1.	2.	3.	4.
5.	6.	7.	8.
9.	10.	11.	12.
13.	14.	15.	16.
17.	18.	19.	20.

Lesson #63

1. What value of x makes the equation true? $3x+12=6x-18$

2. $37+(-12)+41=?$

3. Calculate the circumference of a circle if its diameter is 13 inches.

4. Factor. $12a^3b^2c^2$

5. Solve for x. $x+14=-28$

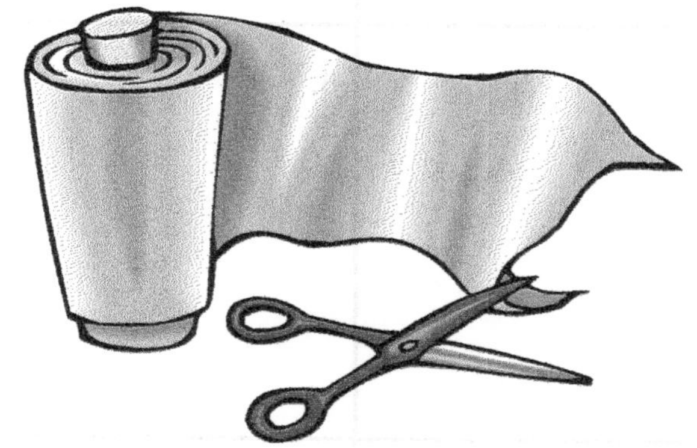

6. $\frac{9}{15} \bigcirc \frac{8}{13}$

7. Find $\frac{3}{4}$ of 24.

8. $5\frac{2}{3}+6\frac{1}{8}=?$

9. A nurse used 20 inches of bandage material to wrap a patient's ankle. A roll of bandage material is 15 feet long. How much of the bandage material was left on the roll?

10. What is the value of x? $\frac{x}{12}=16$

11. $\frac{-108}{9}=?$

12. Write 0.32 as a reduced fraction and as a percent.

13. $-93-(-55)=?$

14. Find the area of the rectangle.

12 mm

8 mm

15. Simplify. $\frac{15x}{21x^2}$

16. Solve for x. $\frac{4}{5}x-\frac{3}{5}x+8=-17$

17. $0.82-0.4365=?$

18. $3(-8)=?$

19. $26.5 \div 0.05=?$

20. Find the GCF of $15x^3y^2$ and $25x^2y^2$.

1.	2.	3.	4.
5.	6.	7.	8.
9.	10.	11.	12.
13.	14.	15.	16.
17.	18.	19.	20.

Lesson #64

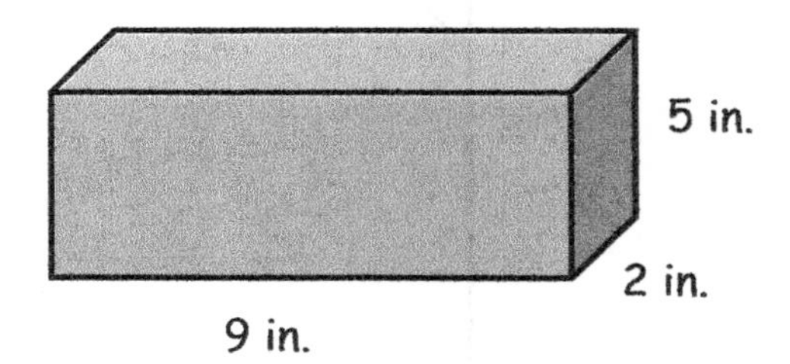

1. $4 \cdot 3 + 8 - 4 + 5^2 = ?$

2. Solve for y. $4y - 2 = 18$

3. Find the surface area of the prism.

4. Find the value of b. $b + 12 = -59$

5. What value of x makes the equation true? $8x + 4 - 3x = 19$

6. $19 - 12\frac{4}{7} = ?$

7. What is the value of a? $\frac{1}{9}a + 8 = 14$

8. Solve for b. $|b - 1| = 6$

9. Write an expression that means *three times a number decreased by 11.*

10. Simplify. $7a + 3b + 5a + 12 + 4b$

11. Find the GCF of $12x^2y^2z^3$ and $24xy^2z^2$.

12. Solve for x. $\frac{x}{15} = 8$

13. Find the value of p. $4(p - 2) + 7p = 14$

14. Put these integers in decreasing order. 15, –13, 27, – 4

15. Solve for y. $-6y = 180$

16. If $a = 3$, $b = 2$ and $c = 4$, what is the value of $a(b + 2c)$?

17. Solve for x. $\frac{x}{3} - 5 = 20$

18. Write 0.8 as a reduced fraction and as a percent.

19. At the end of the first round of the game, Marcus had 23 points. By the end of the second round, his total score was –8. How many points did Marcus lose during the second round?

20. Simplify. $\frac{28a^2b^2}{42a^3b^3}$

1.	2.	3.	4.
5.	6.	7.	8.
9.	10.	11.	12.
13.	14.	15.	16.
17.	18.	19.	20.

Lesson #65

1. Solve for x. $\frac{x}{2}+6=10$

2. Write 45% as a reduced fraction and as a decimal.

3. At the Charity Hospital Race, the last-place runner finished the race at 4:43 p.m. The range of completion times was 2 hours and 36 minutes. If Dennis finished 3 minutes after the first- place runner, at what time did Dennis cross the finish line?

4. $(-9)^3 = ?$

5. $4[3 + 10 \div 5] + 10 = ?$

6. $\begin{pmatrix} -9 & 5 & 0 \\ -6 & 2 & -10 \end{pmatrix} + \begin{pmatrix} -1 & 7 & -8 \\ 4 & -3 & 7 \end{pmatrix} = ?$

7. Solve for a. $8a-2a+12=42$

8. What value of y makes the equation true? $y+9=3y+3$

9. Find the value of x. $2x-8=-16$

10. Find the LCM of $6a^3b^2c^4$ and $14a^2b^2c^2$.

11. $\frac{6}{10}\times\frac{15}{18}=?$

12. $(-3)(-5)(-4) = ?$

13. $0.007 \times 0.06 = ?$

14. $-103 - (-82) = ?$

15. Solve for x. $\frac{1}{9}x+13=24$

16. $55 + (-23) = ?$

17. What is the value of x? $x+4(x+6)=-1$

18. $5\frac{1}{5}-1\frac{4}{5}=?$

19. Solve for x. $\frac{x}{7}=12$

20. Find the missing numbers in the function table.

$y = 2x - 5$

x	y
2	?
3	?
5	?

1.	2.	3.	4.
5.	6.	7.	8.
9.	10.	11.	12.
13.	14.	15.	16.
17.	18.	19.	20.

Lesson #66

1. Find the value of s. $9s+6s=15$

2. Solve for x. $\frac{1}{7}x+14=6$

3. $-100-(-57)=?$

4. What is the value of x? $\frac{x}{4}-2=-7$

5. $25 \div 5 \cdot 3+8 \div 4+2^2=?$

6. Find the value of x. $6x+4(x-3)=8$

7. Rewrite *eight times a number plus twelve* using algebraic symbols.

8. $(-3)(7)=?$

9. Find a^2-b when $a=3$ and $b=2$.

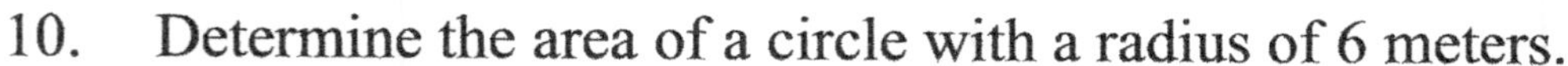

10. Determine the area of a circle with a radius of 6 meters.

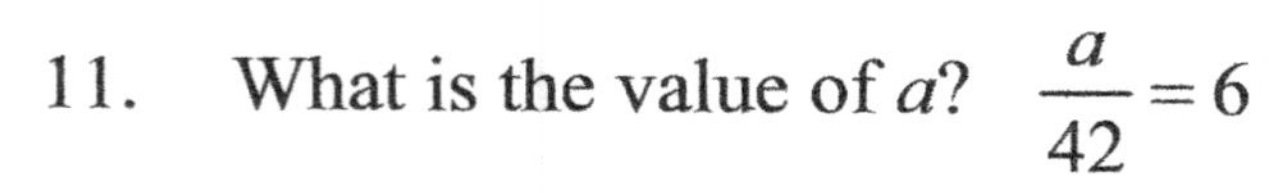

11. What is the value of a? $\frac{a}{42}=6$

12. Solve for y. $4(7+y)=16-2y$

13. Solve for x. $\frac{1}{5}x=13$

14. $66+(-34)=?$

15. Solve for x. $x-7 \leq 15$

16. What is the value of x? $7x=-91$

17. Solve for b. $b+5.6=8.4$

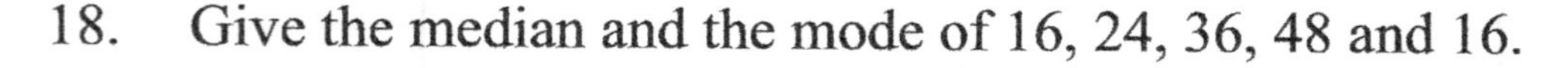

18. Give the median and the mode of 16, 24, 36, 48 and 16.

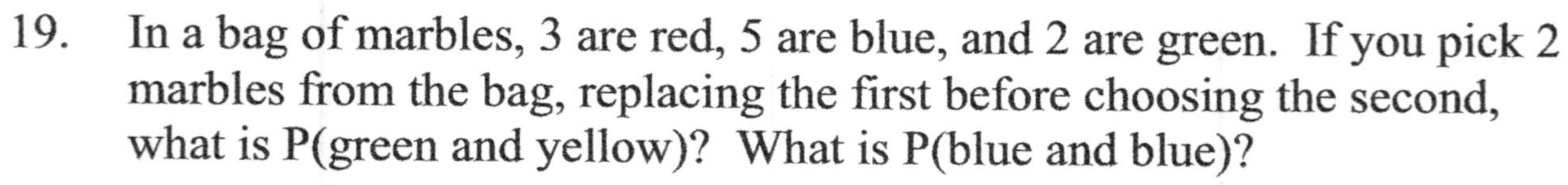

19. In a bag of marbles, 3 are red, 5 are blue, and 2 are green. If you pick 2 marbles from the bag, replacing the first before choosing the second, what is P(green and yellow)? What is P(blue and blue)?

20. Simplify. $\frac{15a^2b^2}{25ab^2}$

1.	2.	3.	4.
5.	6.	7.	8.
9.	10.	11.	12.
13.	14.	15.	16.
17.	18.	19.	20.

Lesson #67

1. $-128-(-56)=?$

2. Solve for a. $a+62=-98$

3. A company's stock went from \$15 a share to \$24 a share. What was the percent of increase in the company stock?

4. Combine like terms. $9a+4ab+6b+3ab+8a$

5. $\frac{7}{8}+\frac{2}{3}=?$

6. When $a=3$ and $b=2$, find a^2+b-5.

7. Solve for x. $4x-12=12$

8. What is the value of x? $\frac{1}{12}x+15=9$

9. $42 \div 6+7 \cdot 2-1+25=?$

10. $-6(-4)(3)=?$

11. Find the value of x. $\frac{x}{5}+13=16$

12. $3 \div 0.05=?$

13. Solve for d. $d-3>10$

14. $215+(-91)=?$

15. Find the measure of each angle.
   a) $\angle ROS$ b) $\angle VOU$

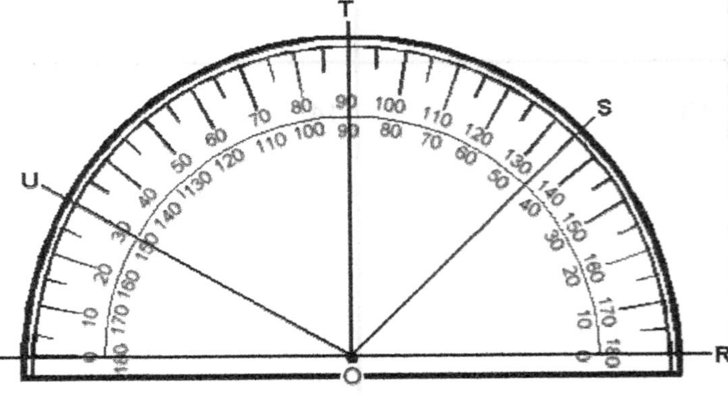

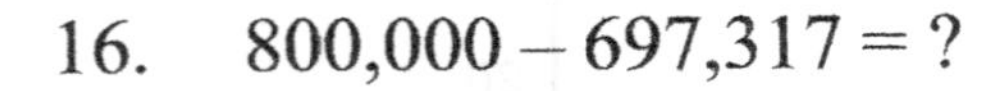

16. $800{,}000-697{,}317=?$

17. Solve for x. $5x+2=7x-2$

18. Round 8.127 to the nearest hundredth.

19. Find the value of a. $6a+3(a+4)=-15$

20. Give the missing numbers for the function shown in the table.

$y=7x+1$

x	y
6	?
3	?
-2	?

1.	2.	3.	4.
5.	6.	7.	8.
9.	10.	11.	12.
13.	14.	15.	16.
17.	18.	19.	20.

Lesson #68

1. What is the value of a? $a+22=69$

2. Find the area of the trapezoid.

10 in.
2 in.
16 in.

3. Solve for x. $6x+3=4x-7$

4. Find the value of x. $\frac{x}{5}=25$

5. $122-(-47)=?$

6. Find the GCF of $12x^2y^3z$ and $18x^3y^2z^2$.

7. Put these integers in increasing order. $-12,\ -26,\ 0,\ 5,\ 10$

8. Write an expression for *the sum of a number and fifteen.*

9. Solve for a. $9a+2(a+7)=-8$

10. $\frac{9}{15}\times\frac{5}{18}=?$

11. $-|-36|=?$

12. $16-3\frac{7}{8}=?$

13. Find the value of x. $\frac{5}{8}x=120$

14. $(-6)(-8)=?$

15. What is the value of x? $3x-5=19$

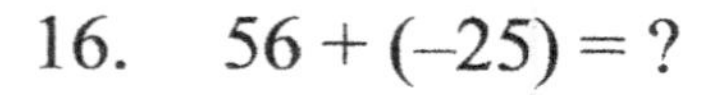

16. $56+(-25)=?$

17. Determine the volume of a rectangular prism with a length of 12 meters, a width of 6 meters, and a height of 3 meters.

18. There are 30 students in Mrs. Jackson's class. One-third of the students wear glasses. Of those students wearing glasses, one-fifth are boys. How many students are boys?

19. $-16+(-24)+13=?$

20. Simplify. $\frac{21a^3b^4c^2}{9a^2b^2c}$

1.	2.	3.	4.
5.	6.	7.	8.
9.	10.	11.	12.
13.	14.	15.	16.
17.	18.	19.	20.

Lesson #69

1. If $x = 4$ and $y = 5$, what is the value of $x + 3y$?
2. $129 - (-73) = ?$
3. Write $\frac{13}{20}$ as a decimal and as a percent.
4. Find the GCF of $8a^3b^5c^2$ and $12a^2b^3c^2$.
5. Solve the equation for b. $b - 56 = 79$
6. $5 + 2[4 + 3(5) - 6] = ?$
7. $-3(-8) = ?$
8. Determine the area of a triangle with a base of 12 inches and a height of 4 inches.
9. $68 + (-22) = ?$
10. $0.7 - 0.4219 = ?$
11. What is the value of x? $\frac{x}{15} = 6$
12. Simplify. $7(3a + 5b - 9)$
13. Solve for x. $x + 37 = -123$
14. Solve for w. $w - 4 < 5$
15. Find the value of x. $4x - 16 = 32$
16. What value of a makes the equation true? $9a + 4 = 14a - 16$
17. Write $6 \cdot 6 \cdot 6 \cdot 6 \cdot 6 \cdot 6$ using a base and an exponent.
18. $0.009 \times 0.004 = ?$
19. Solve for z. $\frac{1}{9}z - 7 = 21$
20. Simplify. $\frac{15x^4y^2z}{20x^3y}$

1.	2.	3.	4.
5.	6.	7.	8.
9.	10.	11.	12.
13.	14.	15.	16.
17.	18.	19.	20.

Lesson #70

1. What is the value of x? $-6x = 90$

2. Solve for x. $\frac{x}{7} = 42$

3. Find the value of a. $\frac{4}{5}a = -20$

4. $\frac{-324}{18} = ?$

5. Solve the equation for x. $x - 19 = 46$

6. $\sqrt{100} - \sqrt{16} + 4^2 = ?$

7. What is the value of x? $6x + 3(x - 8) = 3$

8. During their last vacation, the King family traveled by train. If they traveled 594 miles in 9 hours, what was the average speed of the train?

9. $7\frac{2}{3} + 8\frac{2}{5} = ?$

10. $4^4 = ?$

11. Solve for x. $12x + 3 = 15x$

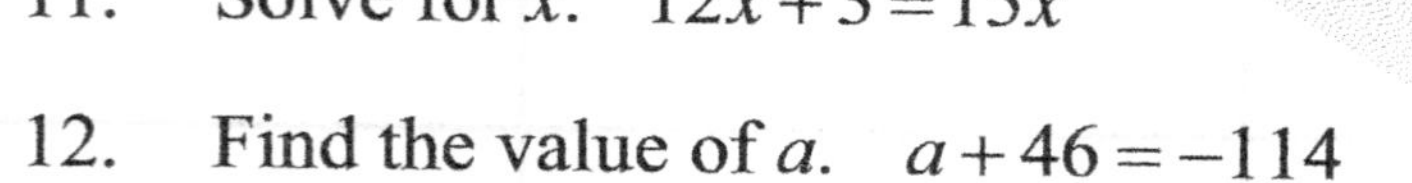

12. Find the value of a. $a + 46 = -114$

13. $8 \cdot 4 + 12 - 16 \div 4 = ?$

14. Evaluate $8b \div b + a$ when $a = 4$ and $b = 5$.

15. Combine like terms. $9a + 4b - 7 + 6a - 2b - 9$

16. $42{,}816{,}755 + 36{,}979{,}327 = ?$

17. Find the GCF of $14a^2b^3c^4$ and $28ab^2c^2$.

18. What is the value of x? $\frac{1}{6}x = 24$

19. Find the missing numbers in the function table.

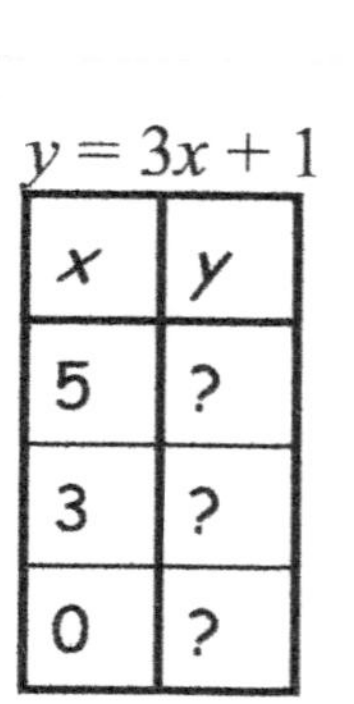

$y = 3x + 1$

x	y
5	?
3	?
0	?

20. Simplify. $\frac{7m^3n^2p}{21m^2n}$

1.	2.	3.	4.
5.	6.	7.	8.
9.	10.	11.	12.
13.	14.	15.	16.
17.	18.	19.	20.

Lesson #71

1. $132 + (-88) = ?$
2. Find the value of x. $3x+9=-18$
3. Solve for x. $\dfrac{x}{9}=10$
4. $4 \cdot 8 \cdot 2 + 9 \div 3 - 6 + 2 = ?$
5. What is the value of x? $\frac{1}{3}x+9=-22$
6. $4x+7xy+9y+8+6xy+3=?$
7. $14 - 9\frac{6}{7} = ?$
8. Solve for c. $c-90=-116$

9. Round 45,688,909 to the nearest million.
10. Find the GCF of $16a^5b^4c^3$ and $24a^4b^2c^4$.
11. What value of x will make the equation true? $10x+17=4x-1$
12. $-7(-4)(-3) = ?$
13. $-66 \bigcirc -45$
14. You have three \$1-bills, two \$5-bills, and a \$20-bill in a bag. If you choose two bills without replacing the 1st one, what is the P(\$1 and \$20)?
15. Find the value of x. $4x-3=5$
16. Solve for b. $b+90=132$
17. If $a=6$ and $b=3$, find $a+3b$.
18. Calculate the area of the square.

15 mm

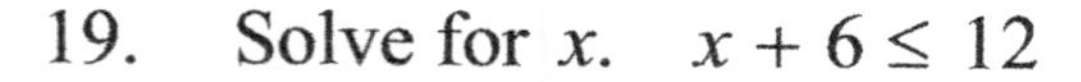

19. Solve for x. $x + 6 \leq 12$
20. Solve for x. $-9x=99$

1.	2.	3.	4.
5.	6.	7.	8.
9.	10.	11.	12.
13.	14.	15.	16.
17.	18.	19.	20.

Lesson #72

1. Find $\frac{3}{5}$ of 40.

2. Solve for x. $\frac{1}{2}x+18=27$

3. Find the value of a. $a+89=-125$

4. What is the value of m? $2m+2=m+7$

5. Solve for x. $2x+3x-4=11$

6. Find the value of x. $\frac{x}{9}=13$

7. $3.4 \times 0.09 = ?$

8. $45 \div 9 \cdot 3 + 10 \div 2 - 4 = ?$

9. What is the value of x? $\frac{1}{7}x=14$

10. Find the value of a. $\frac{1}{5}a-3=0$

11. Write 0.88 as a reduced fraction and as a percent.

12. Solve for h. $h + 7 < -8$

13. Find the missing measurement, x.

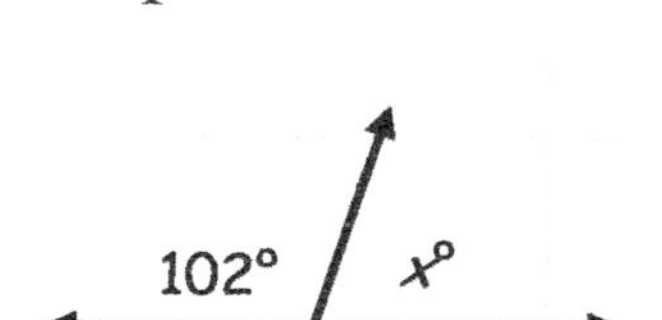

14. 40% of what number is 36?

15. Solve for b. $b-78=-108$

16. $\begin{pmatrix} 8 & 6 \\ -5 & 0 \end{pmatrix} + \begin{pmatrix} -4 & 3 \\ 0 & -2 \end{pmatrix} = ?$

17. Write an algebraic expression that means *a number decreased by 22.*

18. $7{,}000 - 3{,}459 = ?$

19. On the plans for a shed, 1 inch is equal to 4 feet. If the shed is 2 inches tall on the plans, how tall will the actual shed be?

20. $3 \div 0.06 = ?$

1.	2.	3.	4.
5.	6.	7.	8.
9.	10.	11.	12.
13.	14.	15.	16.
17.	18.	19.	20.

Lesson #73

1. $6\frac{2}{3} + 5\frac{1}{5} = ?$

2. Solve for x. $x + 17 = -54$

3. How many feet are in 3 miles?

4. Janet has 36 coins in her purse. One-ninth of the coins are dimes. The value of the dimes is $\frac{1}{4}$ the value of all the coins. How many of Janet's coins are dimes? How much money does Janet have altogether?

5. What value of x makes the fractions equivalent? $\frac{5}{8} = \frac{x}{104}$

6. $3(-4)(2) = ?$

7. Translate the phrase *nine more than two times a number* into an algebraic phrase.

8. $-49 + (-28) = ?$

9. Find the value of y. $3y = -18$

10. When $a = 2$ and $b = 4$, find the value of $a(b + 6)$.

11. Write 48% as a decimal and as a reduced fraction.

12. What is the value of x? $\frac{1}{5}x - 3 = 15$

13. Determine the area of the parallelogram.

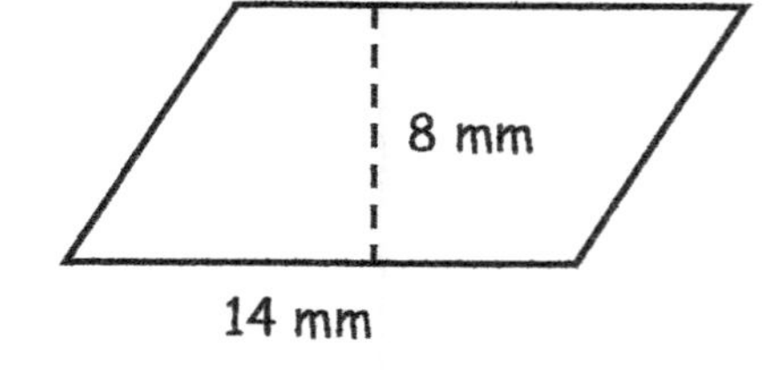

14. Solve for x. $\frac{x}{9} = 12$

15. Distribute. $8(6a + 4b - 7)$

16. What value of x makes the equation true? $2x + 17 = -15$

17. $700{,}000 - 562{,}481 = ?$

18. Combine like terms. $7x + 3y - 8 - 5x + 2y - 9$

19. What is the average of 26, 42, 57, 49 and 16?

20. What type of triangle has exactly two congruent sides?

1.	2.	3.	4.
5.	6.	7.	8.
9.	10.	11.	12.
13.	14.	15.	16.
17.	18.	19.	20.

Lesson #74

1. Write an algebraic expression for *six times a number divided by fourteen.*
2. Solve the equation for x. $x-5=16+4x$
3. What is the range of 16, 98, 32, 44 and 29?
4. Find the value of b. $\frac{1}{6}b+4=-10$
5. How many inches are in 4 yards?
6. Solve for x. $8x-2x+6=-12$
7. Solve for s. $|s-4|=-4$
8. $15-10\frac{3}{7}=?$
9. Round 6.275 to the nearest tenth.
10. $1.36 \div 0.4 = ?$
11. $0.8-0.223=?$
12. Find the value of x. $\frac{x}{9}=14$
13. Write $\frac{6}{25}$ as a decimal and as a percent.
14. $6 \cdot 5 + 4 \div 2 - 2 = ?$
15. Solve for x. $\frac{3}{x}=\frac{7}{28}$
16. Brad made the chart (to the right) to show which sports his classmates like best. If there are 60 students in Brad's class, how many more students chose bicycling than tennis?
17. 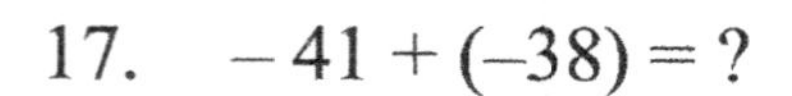 $-41+(-38)=?$
18. Calculate the area of the circle.
19. 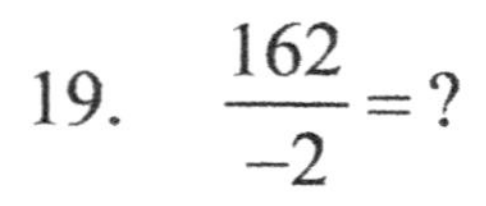 $\frac{162}{-2}=?$
20. Evaluate $(4xy+x)\div 3$ when $x=3$ and $y=2$.

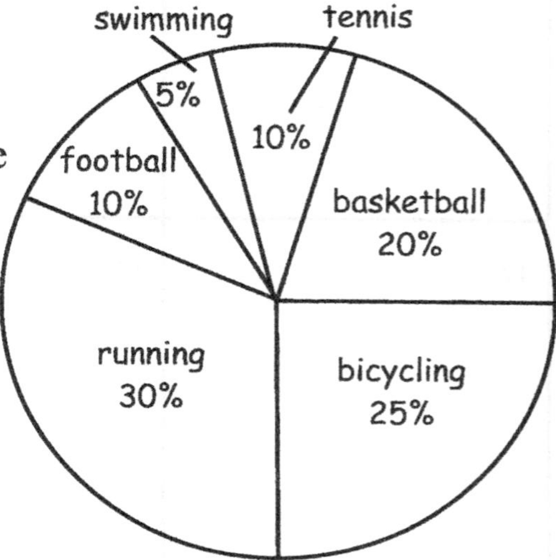

5 in.

1.	2.	3.	4.
5.	6.	7.	8.
9.	10.	11.	12.
13.	14.	15.	16.
17.	18.	19.	20.

Lesson #75

1. What is the area of a triangle with a base of 7 m and a height of 6 m?
2. $\dfrac{-565}{-5} = ?$
3. $45 - 5 \cdot 6 + 7 - 2 = ?$
4. Solve for x. $x + 12 = -44$
5. Simplify. $\dfrac{10x^3y^2}{15xy}$
6. When $x = 6$ and $y = 2$, evaluate $\dfrac{xy}{3} + 5x$.
7. Solve for t. $|t - 3| = 5$
8. Solve for a. $\dfrac{1}{7}a + 3 = 15$
9. Identify the true statement. a) $\dfrac{1}{4} < \dfrac{1}{2}$ b) $\dfrac{1}{4} = \dfrac{1}{2}$ c) $\dfrac{1}{4} > \dfrac{1}{2}$

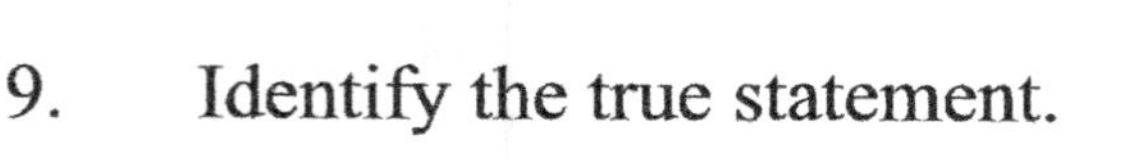

10. $-5(-9) = ?$
11. Find the value of x. $4x - 2 = 10$
12. Combine like terms. $9x - 3y + 4 - 6x - 2y$
13. Round 476,813,241 to the nearest ten million.
14. What is the value of x? $\dfrac{x}{8} + 5 = 11$
15. $33 + (-24) = ?$
16. Calculate the perimeter of the pentagon.

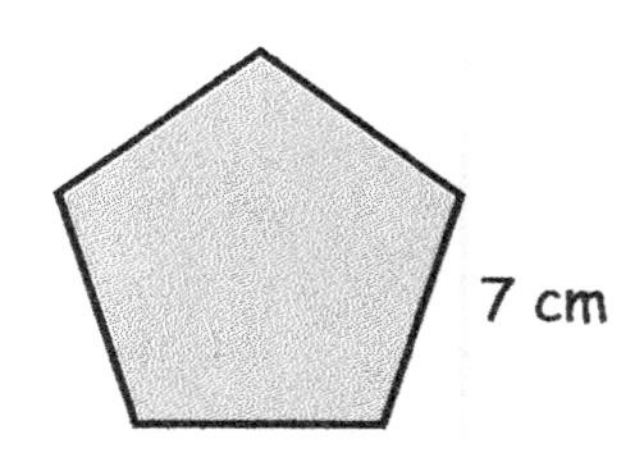

17. Solve for x. $x + 3x - 7 = 29$
18. Write 62% as a decimal and as a reduced fraction.
19. Katie charges \$2.25 per hour for baby-sitting. She is saving everything she earns to buy a new tennis racket which costs \$109.99. Katie has worked a total of 44 hours since she started saving. How much more money does she need to buy the tennis racket?
20. What value of w makes the equation true? $2(w + 8) = 22$

1.	2.	3.	4.
5.	6.	7.	8.
9.	10.	11.	12.
13.	14.	15.	16.
17.	18.	19.	20.

Lesson #76

1. What is the value of x? $14x - 5 = 7 + 10x$

2. $34 - (-18) = ?$

3. Put these integers in decreasing order. $-25,\ 0,\ -36,\ -7,\ -14$

4. $-146 + (-98) = ?$

5. What value of x will make the fractions equivalent? $\frac{12}{44} = \frac{3}{x}$

6. $3(-4)(-2) = ?$

7. 60% of what number is 18?

8. Find the value of $6(x + 2y)$ when $x = 3$ and $y = 2$.

9. Ernie bought four rolls of film for \$15.60. Michelle bought 5 rolls of film for \$16.10. Who got the better value?

10. Solve for x. $\frac{x}{8} = 16$

11. $3[5 + 16 \div 2 - 2] + 5 = ?$

12. $\frac{6}{7} \times \frac{14}{18} = ?$

13. Find the value of x. $-5x = 125$

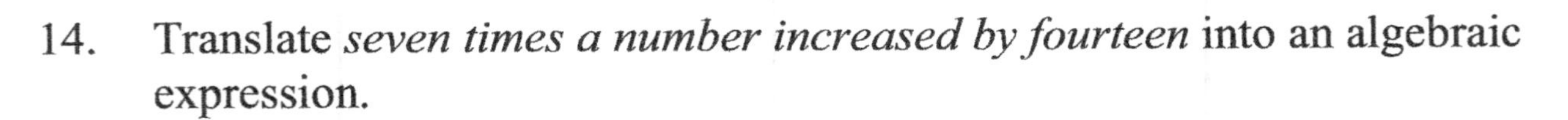

14. Translate *seven times a number increased by fourteen* into an algebraic expression.

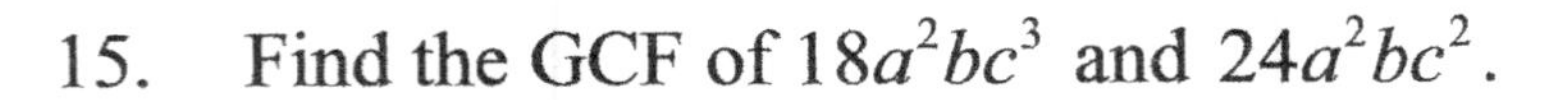

15. Find the GCF of $18a^2bc^3$ and $24a^2bc^2$.

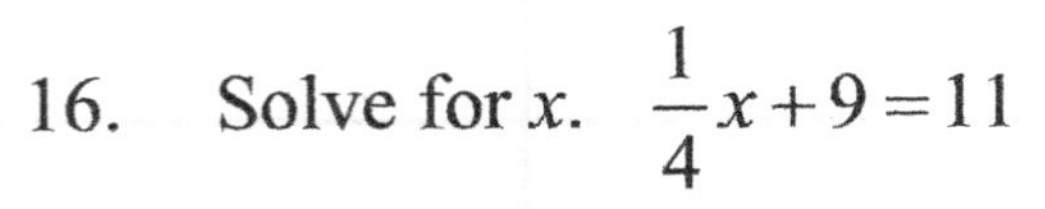

16. Solve for x. $\frac{1}{4}x + 9 = 11$

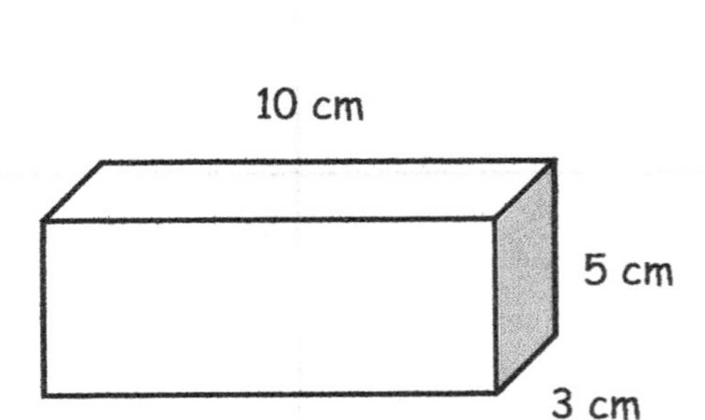

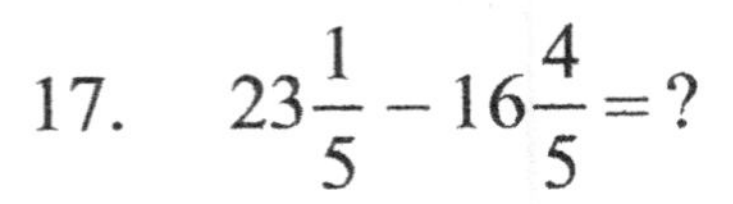

17. $23\frac{1}{5} - 16\frac{4}{5} = ?$

18. Determine the surface area of the rectangular prism above.

19. What is the value of x? $x - 27 = -90$

20. $6.5 + 24.63 = ?$

1.	2.	3.	4.
5.	6.	7.	8.
9.	10.	11.	12.
13.	14.	15.	16.
17.	18.	19.	20.

Lesson #77

1. $835{,}762 + 977{,}349 = ?$

2. $54 + (-26) = ?$

3. What is the value of x? $4x - 7 = 3x + 5$

4. Write 0.5 as a percent and as a reduced fraction.

5. $3.4 \times 0.03 = ?$

6. Solve the equation to find the value of a. $a - 23 = -56$

7. How many centimeters are in 4 meters?

8. What is the value of x? $\frac{x}{7} + 2 = 8$

9. Find the circumference of a circle that has a diameter of 12 cm.

10. $3.36 \div 0.06 = ?$

11. Solve for x. $\frac{x}{4} < 5$

12. 30% of 18 is what number?

13. Solve for b. $-7b = 105$

14. What is the length of each side of a square if its area is 144 ft^2?

15. $3\frac{1}{2} \div 1\frac{1}{4} = ?$

16. $6 + [2(4 + 5) - 3] = ?$

17. Dillon recorded the following numbers: $213 + 1964 + 181 = 2358$, but he accidentally left out the decimal points. Put a decimal point in each number so that the sum equals 235.8.

18. If $r = 2$, $s = 3$ and $t = 6$, find $\frac{rst}{6}$.

19. Which is greater, 0.02 or 12%?

20. $\frac{2x - 1}{5} = 3$ (Hint: Multiply both sides by five.)

1.	2.	3.	4.
5.	6.	7.	8.
9.	10.	11.	12.
13.	14.	15.	16.
17.	18.	19.	20.

Lesson #78

1. What is the value of x? $\frac{x}{8} = -10$

2. A triangle having three sides with different lengths is ____________.

3. Using the x-values in the table, find the corresponding values of y for the function.

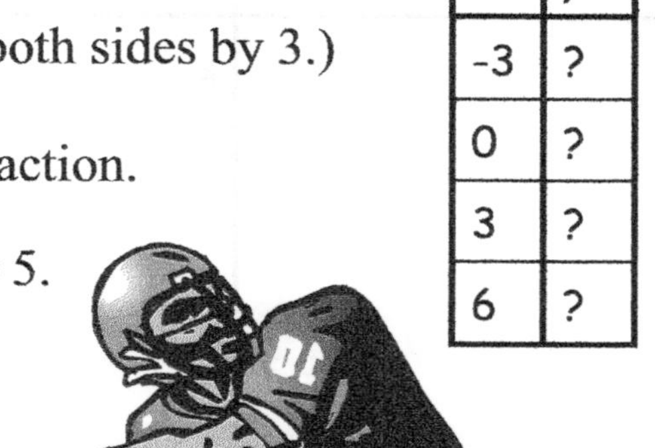

$x + 5 = y$

x	y
-3	?
0	?
3	?
6	?

4. Solve for y. $\frac{y+4}{3} = -1$ (Hint: Multiply both sides by 3.)

5. Write 85% as a decimal and as a reduced fraction.

6. Find $a(b+c) \div 7$ when $a = 3$, $b = 2$ and $c = 5$.

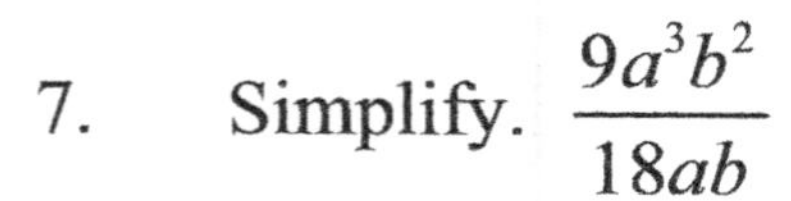

7. Simplify. $\frac{9a^3b^2}{18ab}$

8. How many cups are in 3 pints?

9. Find $\frac{7}{8}$ of 64.

10. $-41 \bigcirc -56$

11. Solve for a. $a + 12 = 44$

12. $6\frac{1}{5} + 3\frac{1}{6} = ?$

13. Find the LCM of $8x^2y^3z$ and $9x^3y^2z^2$.

14. What is the area of a square whose sides are 13 feet long?

15. $-33 - (+16) = ?$

16. Write the phrase using algebra symbols: *The sum of a number and nine.*

17. $24 - 8 - 6 \cdot 2 \div 4 = ?$

18. In a multiplication problem, the answer is the ____________.

19. Distribute. $3x(x-5)$

20. The team's win-loss ratio was 9 to 7. If they won 117 games, how many games were lost?

1.	2.	3.	4.
5.	6.	7.	8.
9.	10.	11.	12.
13.	14.	15.	16.
17.	18.	19.	20.

Lesson #79

1. $86-(-33)=?$
2. Round 813,776,254 to the nearest million.
3. Solve for x. $\dfrac{x+4}{3}=-10$
4. Evaluate $ab-a+b$ if $a=5$ and $b=3$.
5. Find the value of b. $b-12=4b+3$
6. $60-9\cdot 5+21\div 7-3=?$
7. Order these decimals from least to greatest. 6.2 6.02 6.23 6.1
8. $-6(3)(-4)=?$
9. $53+(-21)=?$
10. What is the value of x? $\dfrac{1}{3}x+9=20$
11. Solve for m. $\dfrac{m}{3}>8$
12. Find the LCM of $6a^2b^3$ and $8ab^2$.
13. Distribute. $6x(3x+5)$
14. Solve for a. $5(a-1)=25$
15. **Two angles whose measures total 90° are complementary angles.** What is the measure of the angle that is the complement of the 41° angle?

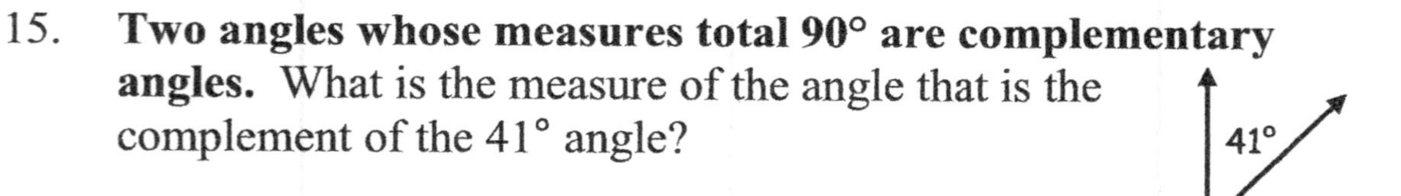

16. Use the protractor below to find the measurement of each angle.

∠AOC ➡ _____ ∠AOE ➡ _____

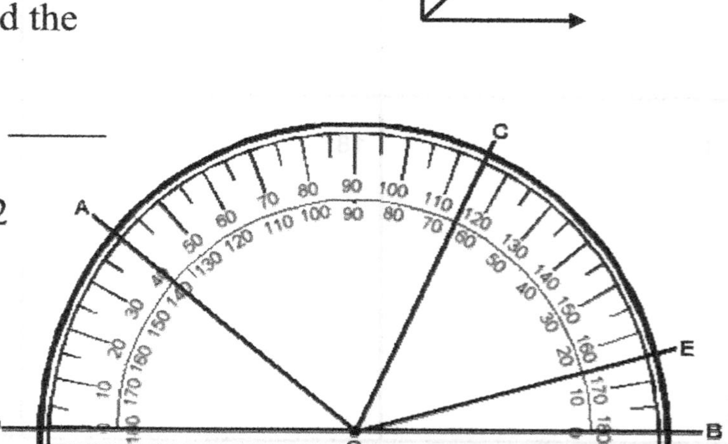

17. Find the value of x. $\dfrac{x}{7}+2=12$
18. $-|-54|=?$
19. Which is greater, 0.7 or 50%?
20. Simplify. $\dfrac{16a^3b^2c}{20a^2b}$

1.	2.	3.	4.
5.	6.	7.	8.
9.	10.	11.	12.
13.	14.	15.	16.
17.	18.	19.	20.

Lesson #80

1. Write 75% as a decimal and as a reduced fraction.

2. $-4(-5)(-3) = ?$

3. Put these integers in decreasing order. $-44,\ 33,\ 0,\ -19,\ 26$

4. What is the value of a? $6a = 90$

5. Solve for x. $2x + 16 = x - 25$

6. Write an expression for *eight more than three times a number.*

7. Evaluate $(3a - b) \div 2$ when $a = 4$ and $b = 2$.

8. Find the LCM of 12 and 14.

9. A straight angle contains how many degrees?

10. Find the GCF of $8x^4y^2$ and $4x^2y$.

11. Find the missing measurement, x.

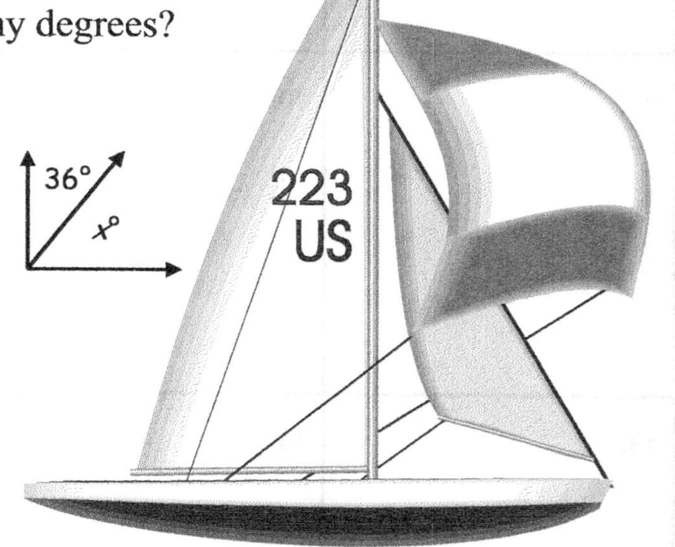

12. Solve for a. $\dfrac{2a + 2}{3} = 8$

13. $125 + (-75) = ?$

14. Find $\dfrac{3}{5}$ of 45.

15. Find the value of a. $\dfrac{1}{3}a - 6 = 5$

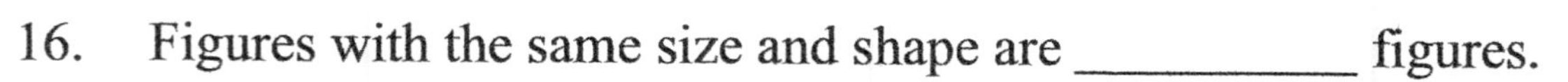

16. Figures with the same size and shape are __________ figures.

17. Determine the area of a parallelogram if it has a base of 14 mm and a height of 3 mm.

18. What is the value of x? $8x - 2x + 3 = 21$

19. Charlie's cat weighs 85 ounces. What is the cat's weight in pounds and ounces?

20. The length of Michael's first boat was 36 feet. His new boat is 47 feet long. What is the percent of increase from his first boat to his new one? Round to the nearest whole number.

1.	2.	3.	4.
5.	6.	7.	8.
9.	10.	11.	12.
13.	14.	15.	16.
17.	18.	19.	20.

Lesson #81

1. Solve for x. $3x+5=-10$
2. Write $\frac{3}{5}$ as a decimal and as a percent.
3. $4 \cdot 4 + 9 \div 3 + 1 - 5 = ?$
4. $\frac{-56}{-7} = ?$
5. A triangle with three congruent sides is a(n) _________ triangle.
6. Simplify. $\frac{15a^3b^2c^2}{25a^2bc^2}$
7. $-65 - (-14) = ?$
8. According to the scale on a map, 1 inch is equal to 30 miles. If two towns are 3 inches apart on the map, how many miles apart are the two towns?

9. What is the value of x? $\frac{x}{4} = 13$
10. Solve for x. $4x+3=2x+9$
11. Give the measure of the angle that is supplementary to the angle below.

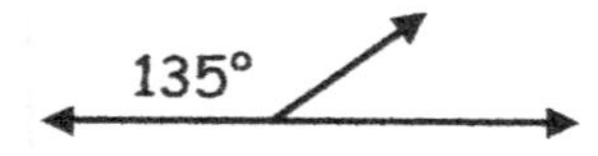

12. Find the value of x. $\frac{7}{8} = \frac{112}{x}$
13. Solve for x. $6x < 36$
14. Which is greater, $\frac{4}{5}$ or 75%?
15. Combine like terms. $7x+4y+9+6x-2y$
16. The answer to an addition problem is the ___________.
17. $90{,}000 - 79{,}366 = ?$
18. If $a=4$ and $b=2$, determine the value of $4(2a-b)$.
19. What percent of 90 is 36?
20. What is the P(2, 4, 3) on 3 consecutive rolls of a die?

1.	2.	3.	4.
5.	6.	7.	8.
9.	10.	11.	12.
13.	14.	15.	16.
17.	18.	19.	20.

Lesson #82

1. $25 - (4 \cdot 3) + 21 \div 7 - 5 = ?$

2. Determine the volume of a rectangular prism whose length is 8 inches, width is 7 inches and height is 3 inches.

3. $-18 + (-12) = ?$

4. If $b = -4$, what is the value of $-5 + b$?

5. How many inches are in 6 yards?

6. Tell whether π is rational or irrational.

7. $\frac{8}{9} \times \frac{12}{16} = ?$

8. In the chart to the right, the measure of the angle formed by Baseball is 140° and the measure of the angle formed by Soccer is 85°. What is the measure of the angle formed by Football?

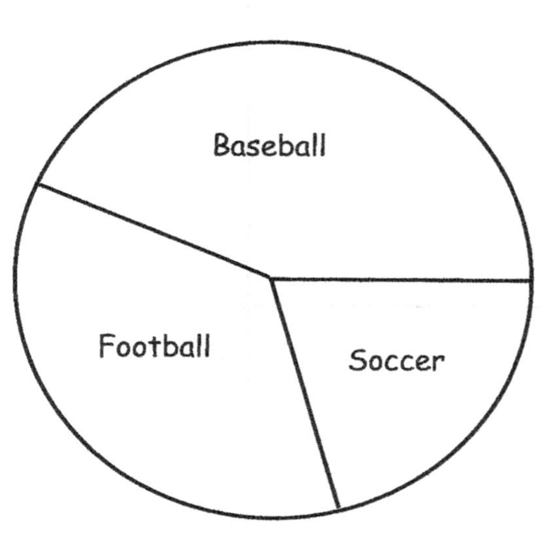

9. $67 - (-24) = ?$

10. Solve for x. $\frac{x}{10} = -15$

11. What is the measure of the missing angle, x?

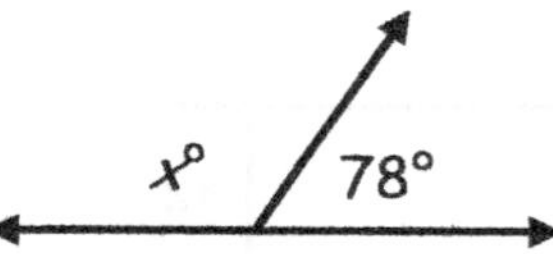

12. Find the value of x. $\frac{1}{4}x = 12$

13. Write an expression that means *eight more than a number.*

14. $-6(7) = ?$

15. Chris is 5 feet 9 inches tall. Give his height in inches.

16. $0.009 \times 0.04 = ?$

17. Solve for x. $10x + 17 = 4x - 1$

18. $37.50 \div 0.75 = ?$

19. Find the value of x. $\frac{x}{5} + 1 = 10$

20. What is the P(H, H, T, T) on 4 flips of a coin?

1.	2.	3.	4.
5.	6.	7.	8.
9.	10.	11.	12.
13.	14.	15.	16.
17.	18.	19.	20.

Lesson #83

1. $31{,}542 - 19{,}695 = ?$

2. $-75 + (-42) = ?$

3. Simplify. $7(3x-6y+4)+2(4x-8y)$

4. Find the value of h. $4h+5=9h$

5. Solve for x. $\frac{8}{12}=\frac{x}{180}$

6. Solve for x. $\frac{1}{8}x-9=22$

7. 60% of 25 is what number?

8. Find the P(T, T, H) on 3 flips of a coin.

9. $6(5)(-3) = ?$

10. Write $\frac{3}{25}$ as a decimal and as a percent.

11. Solve for x. $3x - 7 < 8$

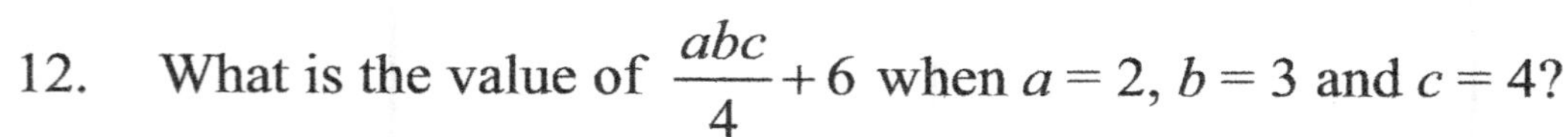

12. What is the value of $\frac{abc}{4}+6$ when $a=2$, $b=3$ and $c=4$?

13. $\begin{pmatrix} -5 & 0 \\ 3 & -9 \end{pmatrix} - \begin{pmatrix} -4 & 7 \\ 2 & -6 \end{pmatrix} = ?$

14. $92 - (-58) = ?$

15. $6.25 + 4.876 = ?$

16. $18\frac{2}{9} - 11\frac{8}{9} = ?$

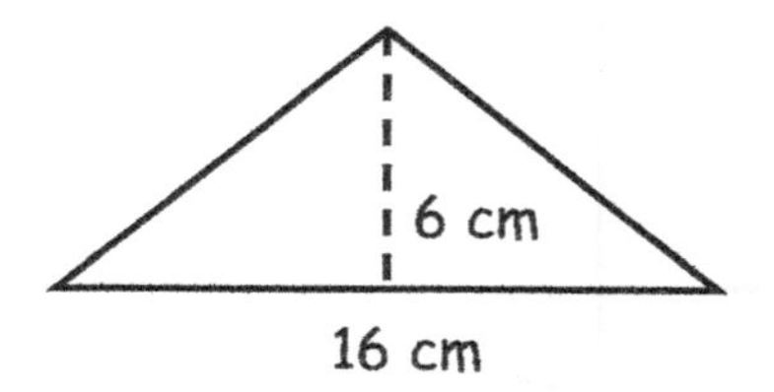

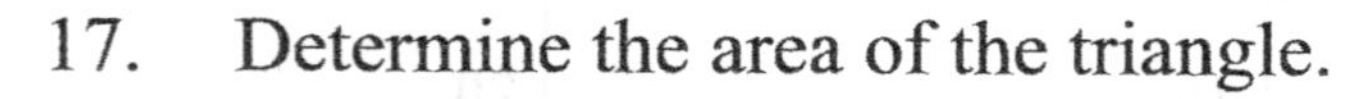

17. Determine the area of the triangle.

18. An office is getting 10 new computers that each need a 3' × 2' work surface. If the office tabletops each measure 4' × 6', how many tables will be needed for all of the computers?

19. Find the LCM of $12x^3y^2z^3$ and $18x^2yz^2$.

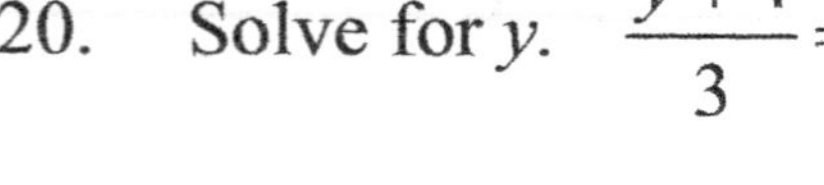

20. Solve for y. $\frac{y+4}{3}=-2$

1.	2.	3.	4.
5.	6.	7.	8.
9.	10.	11.	12.
13.	14.	15.	16.
17.	18.	19.	20.

Lesson #84

1. Order these integers from least to greatest. 34, –62, –28, 7, –3

2. Solve the equation for x. $\frac{8}{9}x-\frac{7}{9}x+4=20$

3. What is the range of 86, 41, 25, 36 and 55?

4. 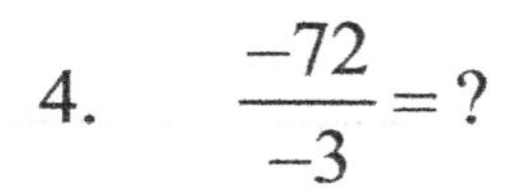 $\frac{-72}{-3}=?$

5. Solve for d. $8d=5d-18$

6. $6 + 3[5 + (8 - 1)] = ?$

7. $6\frac{3}{5}+11\frac{3}{4}=?$

8. $0.42 \times 0.06 = ?$

9. Find the GCF of $9a^3b^3c$ and $12a^2bc$.

10.  Write 0.54 as a reduced fraction and as a percent.

11. Solve to find c. $\frac{2c+1}{7}=3$

12. Solve for a. $2a - 5 \leq 23$

13. What is the value of b? $b-28=-56$

14. $3\frac{1}{5}\times\frac{5}{8}=?$

15. $46 - (-29) = ?$

16. Evaluate $2a+3b-c$ when $a=2$, $b=4$ and $c=3$.

17. How many yards are in a mile?

18. Calculate the area of a circle whose diameter is 18 mm.

19. The ratio of tigers to giraffes in the grasslands is 8 to 10. If there are 104 tigers, how many giraffes are in the grasslands?

20. Translate *sixteen divided by a number decreased by seven* into an algebraic expression.

1.	2.	3.	4.
5.	6.	7.	8.
9.	10.	11.	12.
13.	14.	15.	16.
17.	18.	19.	20.

Lesson #85

1. What is the value of x? $\dfrac{x}{12} = 20$

2. Two angles whose measures add up to 90° are _______ angles.

3. Michael is 6 feet 5 inches tall. What is his height in inches?

4. What percent of 10 is 4? (Hint: Use a proportion.)

5. Marshall's swimming pool is rectangular and holds 4,500 cubic feet of water. How deep is his pool if it is 50 feet long and 20 feet wide?

6. A six-sided polygon is a(n) _______________.

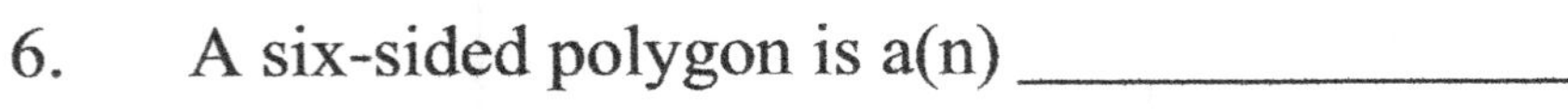

7. Find $\dfrac{2}{9}$ of 27.

8. Draw perpendicular lines.

9. Find the value of b. $b + 19 = 73$

10. $81 \div 9 + 6 \cdot 3 + 21 \div 3 = ?$

11. Solve for x. $-20x = 30 - 5x$

12. How many ounces are the same as 5 pounds?

13. What is the P(5, 1, 3, 1) on 4 rolls of a die?

14. Use the protractor to find the measure of each angle.
∠LON ➡ _____ ∠LOM ➡ _____

15. Solve for x. $\dfrac{x}{5} = \dfrac{36}{60}$

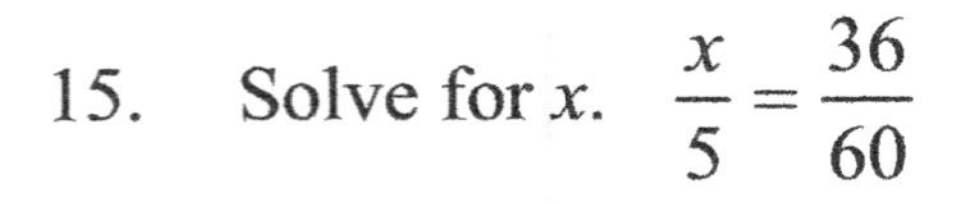

16. $25 \times 36 = ?$

17. Find the value of a. $3a - 7 = 14$

18. What is the value of x? $\dfrac{7x}{3} = -21$

19. Write an expression for a *number increased by 10.*

20. Find the y-values if the domain is {0, 2, –2}. $y = 3x - 2$

1.	2.	3.	4.
5.	6.	7.	8.
9.	10.	11.	12.
13.	14.	15.	16.
17.	18.	19.	20.

Lesson #86

1. Put these decimals in decreasing order. 3.72, 3.072, 3.7, 3.02
2. In a bag of marbles, there are 2 red, 2 blue, and 1 white marble. What is the P(R, B, W) in 3 picks from the bag?
3. What is the value of x? $3x-9=27$
4. A square measures 10 meters on a side. What is its area?
5. $36.2 + 4.865 = ?$
6. $9(-6) = ?$
7. How many millimeters are in a meter?
8. Solve for a. $5a-7=3a+7$
9. A change from 45 meters to 32 meters is what percent of decrease? Round to the nearest whole number.
10. If $a=6$, $b=4$ and $c=3$, find $a(b+c)$.
11. $\frac{9}{15} \div \frac{2}{3} = ?$
12. $\begin{pmatrix} -5 & 5 & 8 \\ -6 & 1 & -3 \end{pmatrix} + \begin{pmatrix} 3 & -4 & 2 \\ 0 & 4 & -3 \end{pmatrix} = ?$
13. $5[6 + 2(3 + 2)] = ?$
14. Solve for y. $2y - 3 > 7$
15. Write 36% as a decimal and as a reduced fraction.
16. $41 + (-18) = ?$
17. What is the range in the function $y = x^2 + 1$ if the domain is $\{3, 0, -2\}$?
18. Find the LCM of $10x^4y^3$ and $25x^2y^2$.
19. What is the value of x? $\frac{x}{8} = 9$
20. Solve for c. $\frac{5c-8}{18} = \frac{2}{3}$ (Hint: Multiply both sides by 18.)

1.	2.	3.	4.
5.	6.	7.	8.
9.	10.	11.	12.
13.	14.	15.	16.
17.	18.	19.	20.

Lesson # 87

1. Find the median of 36, 88, 49, 14 and 23.

Test Scores

Greg's class		Mike's class
8 8	6	2 4
9 6 3	7	0 7 9
7 7 4 0	8	1 4 6
5 5 3	9	1 6 6 7

2. $23\frac{1}{4} + 36\frac{2}{3} = ?$

3. 90% of what number is 36?

4. $8{,}000{,}000 - 657{,}841 = ?$

5. When $a = 4$ and $b = 2$, what is the value of $3(2a + b)$?

6. Combine like terms. $8x - 4y + 7 - 3x - 6y - 2$

7. Use the double stem-and-leaf plot above to decide which class had the better test scores overall.

8. If it is 4:20 p.m. now, what time was it 5 hours ago?

9. Solve for x. $\frac{1}{3}x + 8 = -15$

10. Wayne has a piece of string that is $4\frac{1}{2}$ inches long. He needs to cut the string into 4 equal pieces. What will be the length of each piece?

11. Write an expression for *a number divided by 8 plus 3 times another number.* (Hint: One number could be x and another could be y.)

12. $-|32| = ?$

13. $\sqrt{81} + \sqrt{64} - 2^2 = ?$

14. Find the circumference of a circle if its diameter is 14 meters.

15. What is the value of x? $\frac{9}{7} = \frac{135}{x}$

16. Two angles whose measures add up to 180° are ________ angles.

17. $0.005 \times 0.07 = ?$

18. $17 - 8\frac{6}{7} = ?$

19. Give the name of each shape. a) 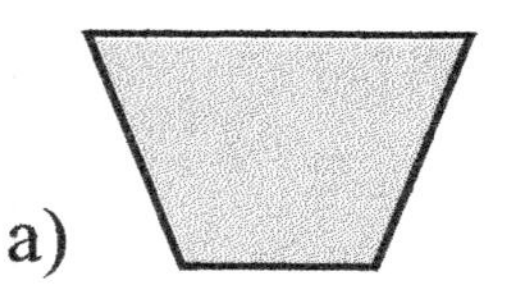 b) 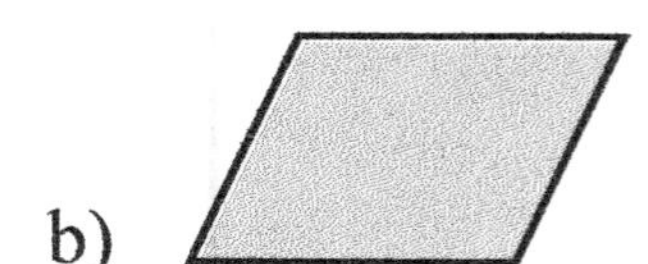

20. Find the y-values in the equation $y = 2x - 1$ when $x = \{-3, 0, 2\}$.

1.	2.	3.	4.
5.	6.	7.	8.
9.	10.	11.	12.
13.	14.	15.	16.
17.	18.	19.	20.

Lesson #88

1. $426 \times 23 = ?$
2. Write $\frac{2}{50}$ as a decimal and as a percent.
3. Find the value of a. $a - 25 = 75$
4. What geometric solid does a can resemble?
5. $-37 - (-12) = ?$
6. How many inches are in 4 yards?
7. Simplify. $\frac{8m^3n^2p}{10mnp}$
8. Mr. Jameson has to repair eight sections of fence. In one hour, he can repair $\frac{2}{3}$ of the sections. How many hours will it take him to repair all eight sections of fence?
9. Solve for x. $8x - 4 = 2x + 20$
10. $18\frac{1}{7} - 9\frac{3}{7} = ?$
11. Determine the volume of the cube.

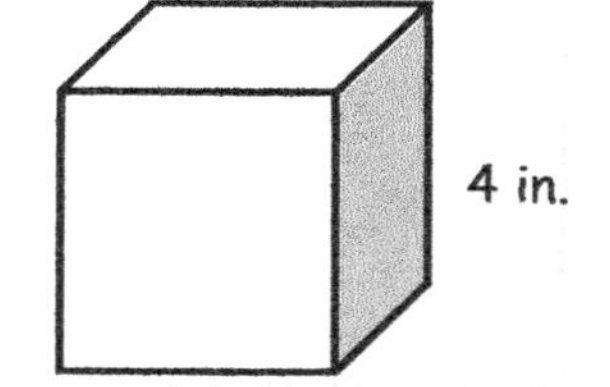

12. Round 56,279,813 to the nearest ten million.
13. Baby Rhia weighs 102 ounces. Give her weight in pounds and ounces.
14. Find the missing measurement, x.

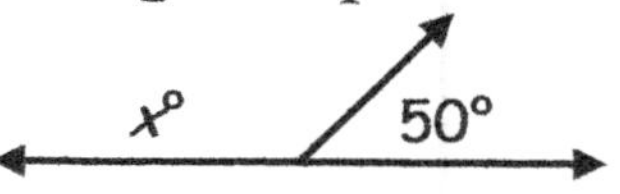

15. What is the P(T, T, T) on 3 flips of a coin?
16. Find the value of x. $\frac{x}{8} = 12$
17. Solve for the value of s. $2(s + 4) \leq 16$
18. When $a = 5$, $b = 4$ and $c = 2$, what is the value of $\frac{ab}{c} - 6$?
19. Solve for x. $6x + 8 = -28$
20. $\frac{-450}{-15} = ?$

1.	2.	3.	4.
5.	6.	7.	8.
9.	10.	11.	12.
13.	14.	15.	16.
17.	18.	19.	20.

Lesson #89

1. Find the value of $x + xy - y$ when $x = 5$ and $y = 3$.

2. *A number decreased by four is three times another number.* Write the sentence as an equation.

3. $\frac{9}{11} \bigcirc \frac{8}{12}$

4. Graph the solution on a number line. $x + 3 > -2$

5. 30% of 18 is what number? (Hint: The answer will be a decimal.)

6. $6 + 2[8 + (5 \cdot 2)] = ?$

7. How many degrees make a straight angle?

8. $-66 - (-42) = ?$

9. $\frac{8}{9} \times \frac{12}{16} = ?$

10. Find the value of b. $b - 27 = -54$

11. Round 8.254 to the nearest tenth.

12. What is the value of x? $7x - 7 = 14$

13. Solve for x. $\frac{4}{3}x = 36$

14. Write the ratio *6:13* in two other ways.

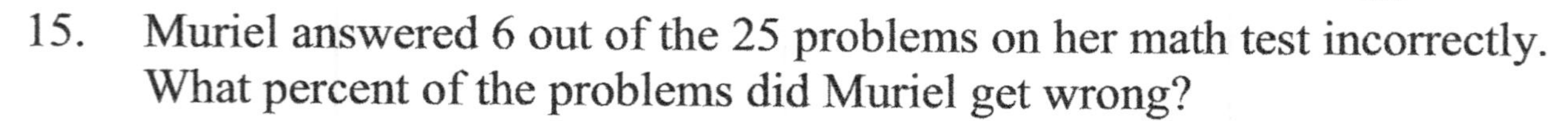

15. Muriel answered 6 out of the 25 problems on her math test incorrectly. What percent of the problems did Muriel get wrong?

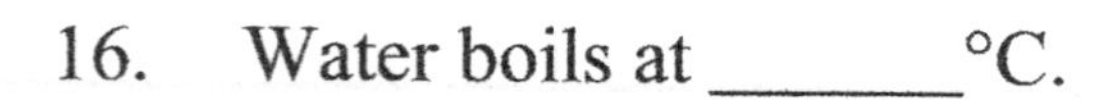

16. Water boils at _______°C.

17. $-14 + (-16) + 8 = ?$

18. Determine the perimeter of a hexagon if its sides each measure seven inches.

19. Find the GCF of $6a^2b^2$ and $18ab^2$.

20. Write the coordinates of points A, C, E, and F.

1.	2.	3.	4.
5.	6.	7.	8.
9.	10.	11.	12.
13.	14.	15.	16.
17.	18.	19.	20.

Lesson #90

1. $-7(9) = ?$

2. Find the range of 86, 24, 100, 16 and 12.

3. $65 + (-41) = ?$

4. A band marches approximately $2\frac{1}{4}$ miles at each practice. If the band practices 4 times a week, how many miles will the members have marched in 4 weeks?

5. $19\frac{3}{7} - 12\frac{6}{7} = ?$

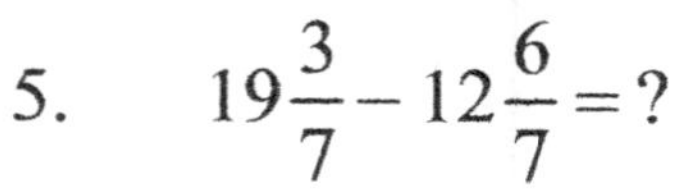

6. What is the value of x? $\frac{3}{4}x = 24$

7. How many feet are in 7 yards?

8. Find the GCF of $10x^2y^2z^2$ and $12xy^2z$.

9. Solve for a. $a + 24 = 63$

10. Write $\frac{3}{25}$ as a decimal and as a percent.

11. What value of y makes the equation true? $6y - 9 = 3y + 18$

12. Solve and graph the solution on a number line. $w + 2 > -1$

13. $63 \div 9 + 8 \cdot 3 - 6 = ?$

14. Solve for x. $\frac{x}{7} = 14$

15. On the Fahrenheit scale, water boils at ____________.

16. Solve for x. $7x + 4 = -17$

17. When $x = \{3, 0, -1\}$, find the corresponding y-values. $y = 4x$

18. Rewrite *six times a number divided by nine* using algebraic symbols.

19. Simplify. $\frac{12a^3bc^2}{18ac}$

20. Find the area of the parallelogram.

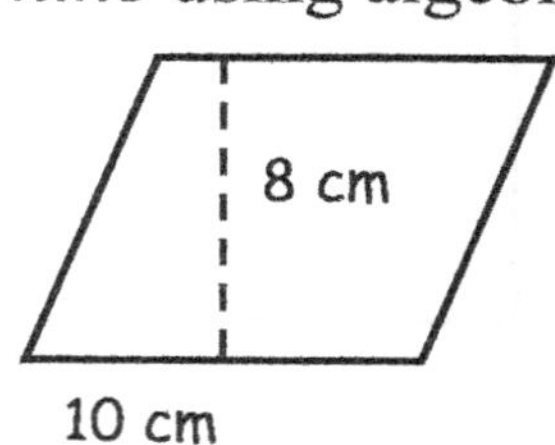

1.	2.	3.	4.
5.	6.	7.	8.
9.	10.	11.	12.
13.	14.	15.	16.
17.	18.	19.	20.

Lesson #91

1. If $a = 3$, $b = 8$ and $c = 2$, find the value of $a(b \div 2) + c$.

2. $36 - (-10) = ?$

3. $5{,}000{,}000 - 979{,}816 = ?$

4. Solve the inequality for w. $w + 4 \leq 8$

5. Round 37,846,552 to the nearest hundred thousand.

6. What value of x makes the fractions equivalent? $\frac{x}{9} = \frac{63}{81}$

7. Distribute and combine like terms. $3(4a - 7) + 5(2a - 3b - 3)$

8. $\frac{84}{-7} = ?$

9. Solve for x. $\frac{x}{9} = 5$

10. Two angles whose measures add up to 90° are ________ angles.

11. Find the value of a. $7a - 4 = 2a - 24$

12. One-fourth of Evelyn's 12 friends want pizza, and one-half of them want tacos. The rest of her friends are not sure what they want. How many of Evelyn's friends are not sure?

13. Write 65% as a decimal and as a reduced fraction.

14. A rectangular prism has a length of 5 meters, a width of 4 meters, and a height of 3 meters. Calculate its surface area.

15. Find the missing measurement, x.

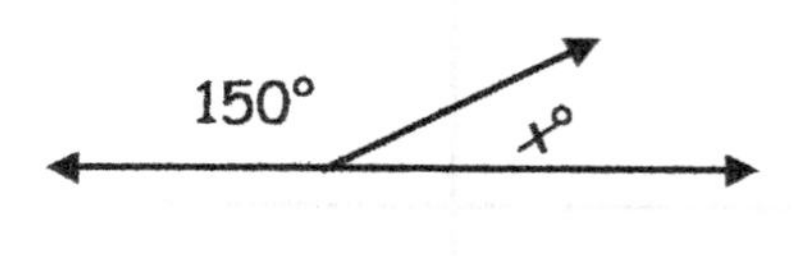

16. Solve for a. $\frac{1}{8}a - 7 = 13$

17. $\frac{14}{21} \times \frac{7}{2} = ?$

18. Use the table to find the y-values in the equation.

$y = -7x + 1$

x	y
-4	?
0	?
-5	?

19. Find the LCM of $8x^2y^3z$ and $15xy^2z^2$.

20. Solve for b. $b - 24 = -80$

1.	2.	3.	4.
5.	6.	7.	8.
9.	10.	11.	12.
13.	14.	15.	16.
17.	18.	19.	20.

Lesson #92

1. $378{,}884 + 853{,}762 = ?$

2. Solve for a. $10a - 5a + 4 = -21$

3. What are the corresponding y-values in the equation $y = 2x + 3$ when $x = \{2, 0, -1\}$?

4. Graph the solution on a number line. $-5 > b - 1$

5. $0.8 - 0.3264 = ?$

6. Find the value of x. $\frac{1}{7}x - 3 = 25$

7. What is the probability that 3 and 1 will come up on 2 rolls of a die?

8. Solve for x. $3x - 8 = 28$

9. Write the formula for finding the surface area of a rectangular prism.

10. How many yards are in 2 miles?

11. $-6(-3)(2) = ?$

12. 40% of 25 is what number?

13. What is the value of x? $6x = 4(x + 5)$

14. Rewrite the phrase as an algebraic expression: *Two times a number increased by nine times another number.*

15. $5(9 + 2) - 5 = ?$

16. On a Celsius thermometer, water freezes at __________.

17. On Saturday, Henry spent $1\frac{1}{2}$ hours cutting the grass, $1\frac{3}{4}$ hours pulling weeds, and 2 hours sweeping and bagging the weeds. In all, how much time did Henry spend on yard work?

18. $0.004 \times 0.003 = ?$

19. $\frac{8}{10} \div \frac{2}{10} = ?$

20. Simplify. $\frac{8mn^2p^3}{12m^2n^2p}$

1.	2.	3.	4.
5.	6.	7.	8.
9.	10.	11.	12.
13.	14.	15.	16.
17.	18.	19.	20.

Lesson #93

1. What is the value of x? $\frac{x}{8}=14$

2. Find the area of a circle if its radius is 6 millimeters.

3. $-33+(-17)=?$

4. Graph the solution on a number line. $2<s-8$

5. $24\frac{1}{5}+13\frac{3}{4}=?$

6. How many quarts are in 8 gallons?

7. Write $\frac{5}{8}$ as a decimal.

8. Evaluate $\frac{ab}{3}-4$ when $a=6$ and $b=4$.

9. Solve for a. $7a+6=3a-22$

10. $\frac{-81}{3}=?$

11. $-|-16|=?$

12. What is the P(1, 2) on 2 spins?

13. Find the value of x. $\frac{8}{12}=\frac{x}{144}$

14. $4.26+35.7=?$

15. A triangle with 3 congruent sides is a(n) ___________ triangle.

16. Solve for a. $a-24=-56$

17. $6[3(4+3)-5]+2^2=?$

18. **A complex shape can be divided into two rectangles.** Find the area of each one, and then add.

19. Simplify. $\frac{7a^3b^2c^3}{21a^2bc}$

20. Calculate the area of the trapezoid.

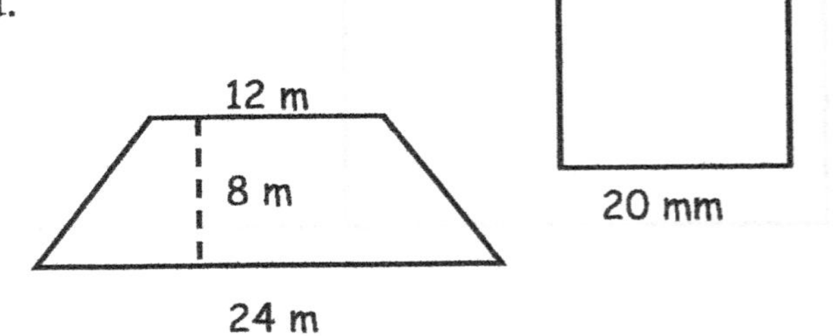

1.	2.	3.	4.
5.	6.	7.	8.
9.	10.	11.	12.
13.	14.	15.	16.
17.	18.	19.	20.

Lesson #94

1. How many inches are in 4 yards?

2. $-77 + (-86) = ?$

3. Tell whether $\sqrt{121}$ is rational or irrational.

4. $\begin{pmatrix} -12 & 6 \\ 11 & -9 \end{pmatrix} - \begin{pmatrix} -7 & 8 \\ 3 & -6 \end{pmatrix} = ?$

5. What is the P(T, T, H) on 3 flips of a coin?

6. $-6(3)(-4) = ?$

7. Find the GCF of $15a^3b^3c^2$ and $20a^2bc$.

8. Solve for x. $7x - 4 = 3x + 24$

9. Find the area of the trapezoid.

4 ft. 9 ft. 10 ft.

10. $20 - 13\frac{4}{5} = ?$

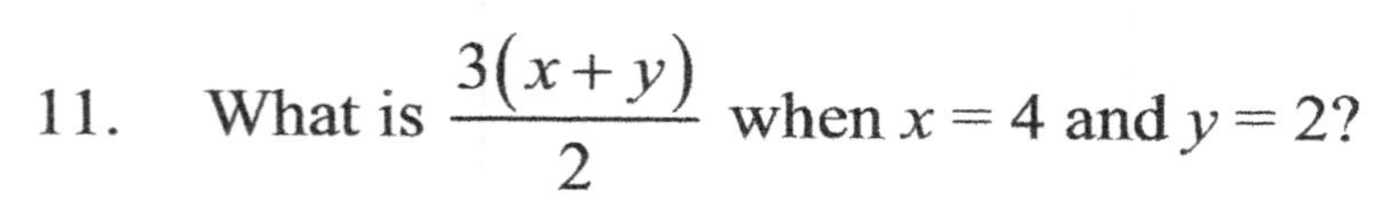

11. What is $\frac{3(x+y)}{2}$ when $x = 4$ and $y = 2$?

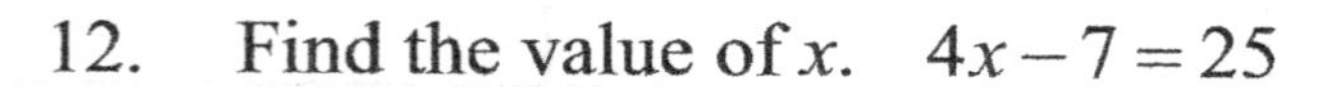

12. Find the value of x. $4x - 7 = 25$

13. Water boils at ______° Fahrenheit.

14. Solve the inequality for x. $x - 1 > 10$

15. When $x = \{2, 0, -3\}$, find the values for y. $y = 4x - 3$

16. Find the median and the mode of 72, 56, 23, 88 and 56.

17. Solve for x. $\frac{x}{14} = 13$

18. $5 + 7 \cdot 5 - 36 \div 6 + 3 = ?$

19. Determine the circumference of a circle with a diameter of 12 cm.

20. Order these integers from greatest to least.

    5 −12 0 −27 −1

1.	2.	3.	4.
5.	6.	7.	8.
9.	10.	11.	12.
13.	14.	15.	16.
17.	18.	19.	20.

Lesson #95

1. $92-(-35)=?$
2. Find the area of this trapezoid.

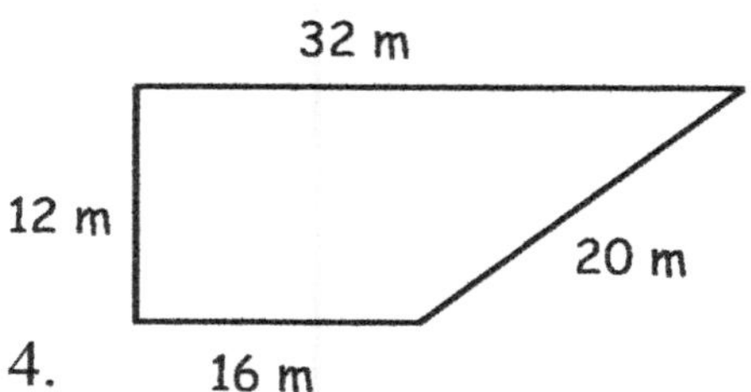

3. Solve for b. $b+22=70$
4. Evaluate $\frac{ab+c}{5}$ when $a=3$, $b=2$ and $c=4$.
5. Graph the solution on a number line. $7+a<1$
6. $16\frac{3}{7}+10\frac{2}{5}=?$

7. $63 \div 7 + 8 \cdot 4 - 10 + 1 = ?$
8. How many feet are in 2 miles?
9. $\frac{-81}{-9}=?$
10. $0.6-0.2133=?$
11. Solve for x. $\frac{1}{9}x+6=10$
12. Solve the compound inequality for n. $-6<3n+9<21$
13. Simplify. $\frac{6x^3y^2z}{12xyz^2}$
14. What is the value of x? $10x-5=4x+25$
15. Write *eight divided by a number increased by twelve* in algebraic symbols.
16. $120+(-76)=?$
17. Sylvia has 5 classes before lunch: language, social studies, math, art, and science. Language comes after art, but before social studies. Sylvia goes to lunch right after science and is usually late to math class from language class. What is Sylvia's schedule before lunch?
18. $\frac{12}{15}\times\frac{10}{18}=?$
19. Write the formula for finding the area of a triangle.
20. Write 0.22 as a reduced fraction and as a percent.

1.	2.	3.	4.
5.	6.	7.	8.
9.	10.	11.	12.
13.	14.	15.	16.
17.	18.	19.	20.

Lesson #96

1. What is x when the fractions are equivalent? $\frac{9}{11}=\frac{x}{121}$

2. $-7(-2)(-3)=?$

3. Find the value of a. $a-6.7=9.2$

4. 25% of 60 is what number?

5. Solve for x. $2 < x + 5 < 9$

6. $45-36\frac{3}{7}=?$

7. Solve for x. $\frac{4}{5}x=100$

8. $80{,}000-32{,}466=?$

9. How many centimeters are in 7 meters?

10. Find the missing measurement, x.

11. What is the value of a? $4a=-120$

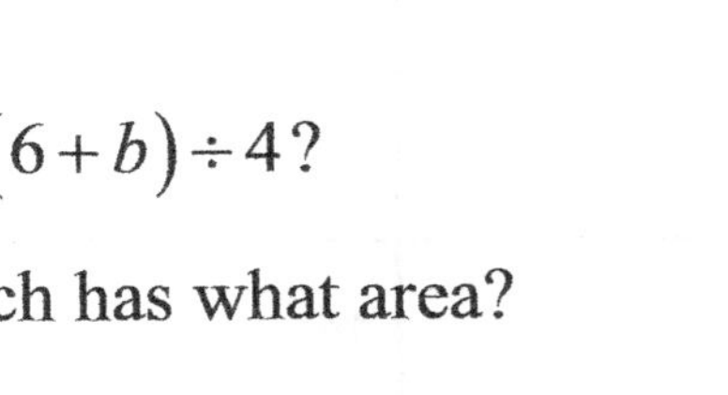

12. If $a=2$ and $b=6$, what is the value of $a(6+b)\div 4$?

13. A square whose sides measure 12 feet each has what area?

14. $120+(-78)=?$

15. What is the P(1, 5, 2) on 3 spins?

16. Draw two congruent pentagons.

17. Find the value of a. $8a+9=2a-15$

18. $\frac{8}{10}\div\frac{4}{10}=?$

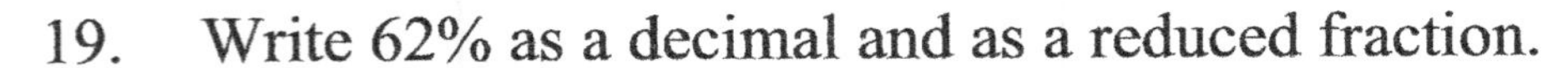

19. Write 62% as a decimal and as a reduced fraction.

20. The ratio of trucks to cars on the turnpike was 8 to 9. If there were 96 trucks on the turnpike, how many cars were there?

1.	2.	3.	4.
5.	6.	7.	8.
9.	10.	11.	12.
13.	14.	15.	16.
17.	18.	19.	20.

Lesson #97

1. What is the value of a? $\frac{1}{3}a - 8 = 12$
2. $16\frac{2}{7} - 10\frac{5}{7} = ?$
3. $2 + 3[5 + 2(4 + 5)] = ?$
4. Find the GCF of $8x^3y^2z$ and $12xyz$.
5. $-49 + (-63) = ?$
6. Solve for x. $x - 52 = 125$
7. The average weight of three colts is 298 pounds. Two of the colts weigh 255 and 260 pounds each. What is the weight of the third colt?
8. What value of x makes the equation true? $9x - 8 = 3x + 10$
9. Find the value of u. $u - 8 = -15$
10. When $a = 3$, $b = 5$ and $c = 2$, find $\frac{abc}{6} + 3$.
11. Find the y-values in the equation $y = x - 7$ when $x = \{0, -5, 8\}$.
12. $-8(3)(-2) = ?$
13. $1.52 \times 0.03 = ?$
14. Write $\frac{3}{25}$ as a decimal and as a percent.
15. Solve for x. $x > 2x + 9$
16. What is the value of x? $6x = -72$
17. $\begin{pmatrix} -9 & 4 & 0 \\ 6 & -1 & 7 \end{pmatrix} - \begin{pmatrix} -4 & 1 & 3 \\ 3 & -8 & -2 \end{pmatrix} = ?$
18. 80% of 35 is what number?
19. Simplify. $3(4a + 7b - 5) + 2(3a - 4)$
20. Simplify. $\frac{18a^3b^2c^4}{24a^2bc^2}$

1.	2.	3.	4.
5.	6.	7.	8.
9.	10.	11.	12.
13.	14.	15.	16.
17.	18.	19.	20.

Lesson #98

1. What is the volume of a rectangular prism if its length is 16 mm, its width is 4 mm, and its height is 3 mm?

2. $18\frac{3}{7}+9\frac{1}{14}=?$

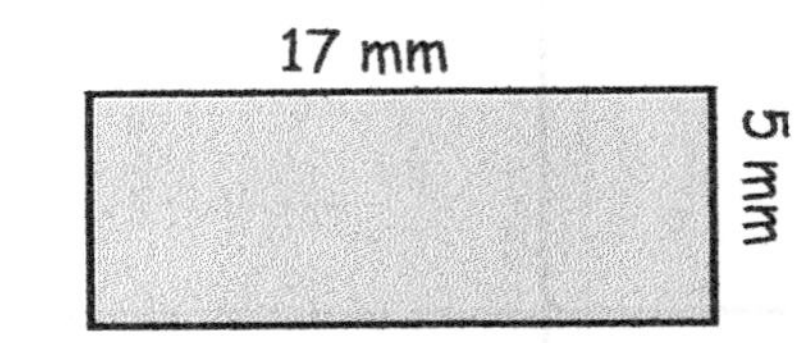

3. Determine the area of the rectangle.

4. What is the value of b? $b-32=77$

5. Write an expression that means *a number, m, increased by 150.*

6. Solve for x. $\frac{2}{3}x=120$

7. $37.2-8.65=?$

8. Combine like terms. $3a-4b+8-2a-5b+2$

9. Find the value of x. $\frac{1}{5}x+3=15$

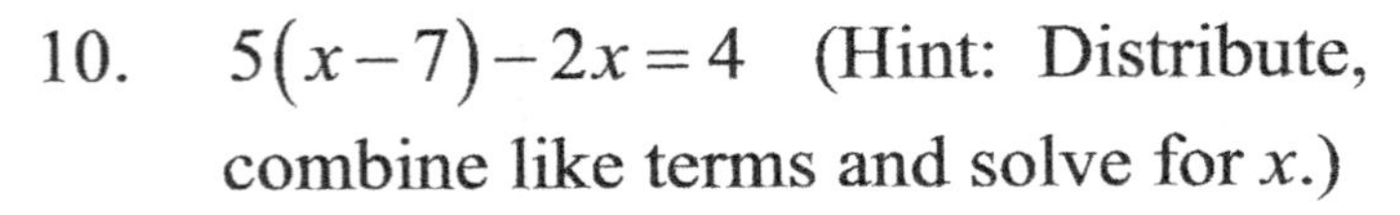

10. $5(x-7)-2x=4$ (Hint: Distribute, combine like terms and solve for x.)

11. Write 0.04 as a percent and as a reduced fraction.

12. Solve the compound inequality. $6<x+5\leq 11$

13. What is the value of $\frac{abc}{5}+10$ when $a=2$, $b=3$ and $c=10$?

14. Solve for x. $\frac{9}{12}=\frac{x}{120}$

15. Find the value of x. $7x+4=x+28$

16. Find the LCM of $8x^3y^2z$ and $10x^2yz^2$.

17. Write 6.0073 using words.

18. Solve the inequality for x. $x+6\leq 7$

19. Find the value of x. $3x-9=18$

20. Give the coordinates of each point.

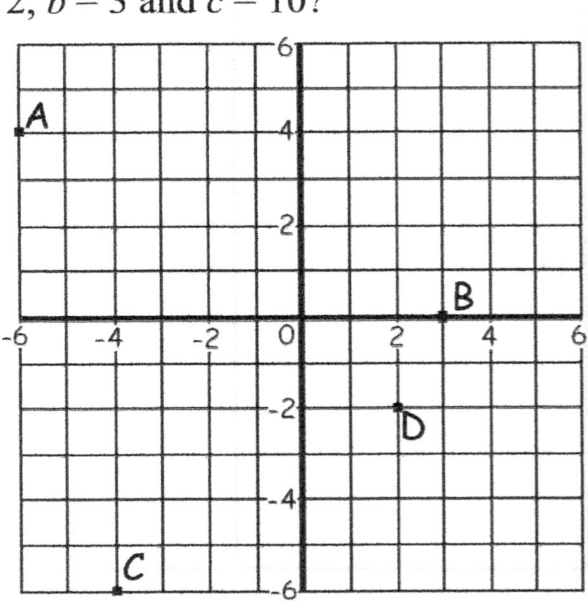

1.	2.	3.	4.
5.	6.	7.	8.
9.	10.	11.	12.
13.	14.	15.	16.
17.	18.	19.	20.

Lesson #99

1. Find the median and the mode of 16, 39, 82, 24 and 82.
2. 300,000 – 198,654 = ?
3. Write $\frac{9}{25}$ as a decimal and as a percent.
4. Solve the compound inequality. $2n + 3 < 7 \text{ or } -n + 9 \leq 2$
5. –155 – (– 46) = ?
6. What is the value of x? $\frac{4}{5}x = 40$
7. Put these units of volume in increasing order. pint, quart, cup, gallon
8. When $x = 3$ and $y = 2$, what is the value of $2x + 4y - 7$?
9. Write the ratio $\frac{5}{9}$ two other ways.
10. Find the area of the trapezoid.

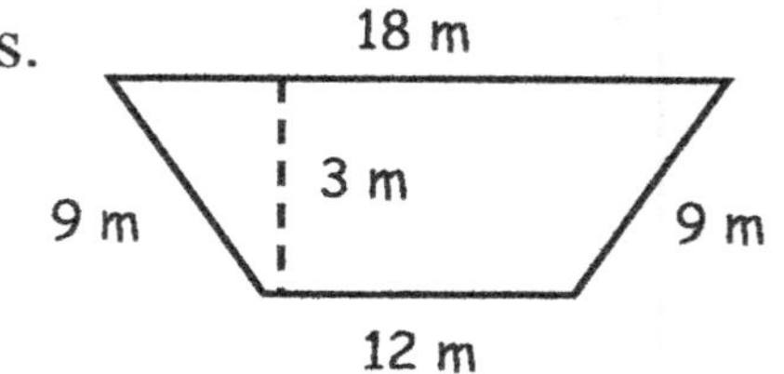

11. 6 + 4[8 + 3(10 – 5)] = ?
12. –3(5)(2) = ?
13. Find the value of a. $\frac{1}{10}a - 8 = 24$
14. –13 + (–15) + 9 = ?
15. Solve for x. $12x - 4 = 3x + 32$
16. At the nature preserve, the ratio of kangaroos to lions is 8 to 9. If there are 117 lions, how many kangaroos are there?

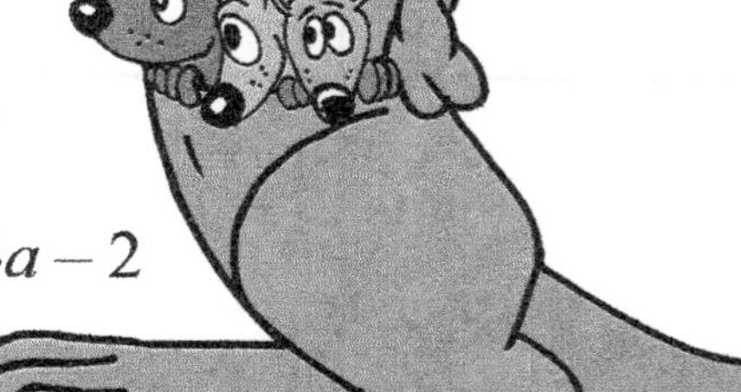

17. Combine like terms. $8a - 4b + 6 - 3b + 4a - 2$
18. How many cups are in 4 pints?
19. A triangle with no congruent sides is called ____________.
20. Order these integers from greatest to least. –29, 0, –7, –16, –4

1.	2.	3.	4.
5.	6.	7.	8.
9.	10.	11.	12.
13.	14.	15.	16.
17.	18.	19.	20.

Lesson #100

1. How many inches are in 4 yards?

2. $362{,}475{,}886 + 869{,}753{,}458 = ?$

3. $-19 - (-9) = ?$

4. Addition Rule of Integers: When the signs are the same, __________ and bring down the sign. When the signs are different, __________ and take the sign of the ________ number.

5. What is the area of a triangle if its base is 16 cm and its height is 4 cm.

6. $|-32| = ?$

7. Find the missing measurement, x.

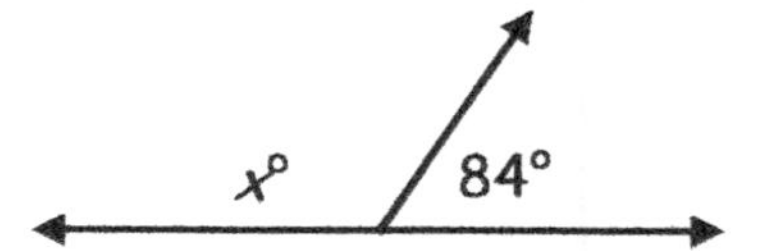

8. Solve for x. $7x + 4 = -17$

9. Write an expression for *a number divided by 3, decreased by 9.*

10. $18\frac{2}{7} - 12\frac{6}{7} = ?$

11. On the spinner, what is the P(8, a number greater than 2) on two spins?

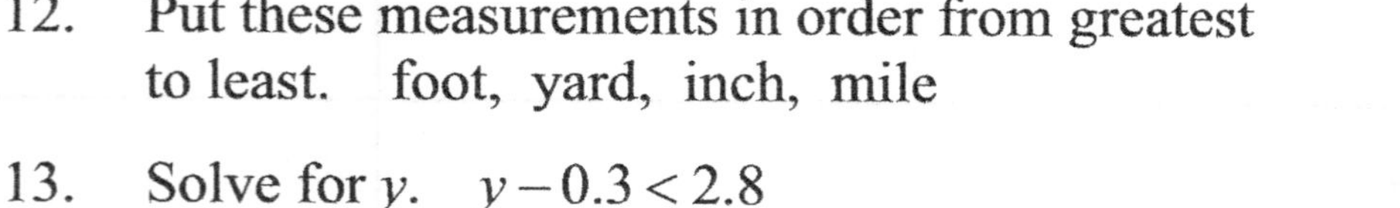

12. Put these measurements in order from greatest to least. foot, yard, inch, mile

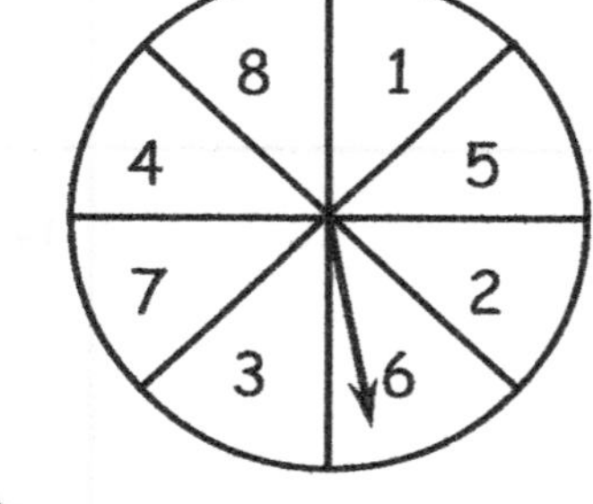

13. Solve for y. $y - 0.3 < 2.8$

14. $4 \div 0.08 = ?$

15. $3\frac{1}{2} \times 4\frac{2}{5} = ?$

16. $6\frac{3}{7} + 4\frac{2}{5} = ?$

17. If $x = 6$ and $y = 9$, find $\frac{x+y}{3} + 10$.

18. How many pounds are in 6 tons?

19. On the Celsius scale, what is water's boiling point?

20. Use the table to find the y-values in the equation, $y = \frac{x}{2} + 3$.

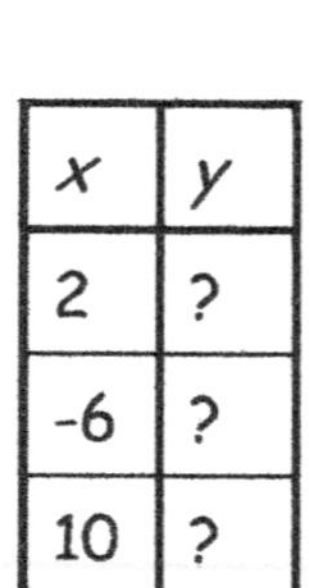

x	y
2	?
-6	?
10	?

1.	2.	3.	4.
5.	6.	7.	8.
9.	10.	11.	12.
13.	14.	15.	16.
17.	18.	19.	20.

Lesson #101

1. What is the value of x? $\dfrac{x}{13}=10$

2. Two angles whose measures add up to 180° are __________ angles.

3. Find the value of b. $b-27=-56$

4. $10[6+4(3+2)-1]=?$

5. Use algebraic symbols to express *three times a number, divided by twelve.*

6. Write $\dfrac{1}{8}$ as a decimal.

7. $66+(-27)=?$

8. 60% of 80 is what number?

9. Determine the circumference of a circle with a diameter of 8 meters.

10. 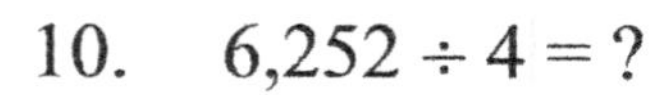 $6{,}252 \div 4 = ?$

11. $\dfrac{-125}{5}=?$

12. Evaluate $x+xy-y$ when $x=4$ and $y=2$.

13. Solve for d. $4d \leq -28$

14. The frame of a new house was 15 feet tall. After the roof was added, it was 25 feet tall. What was the percent of change in the height of the new house? Round to the nearest whole number.

15. What is the value of c? $3(c+4)=15$

16. $2^6=?$

17. Find the GCF of $6a^3b^2c$ and $12a^2b^2c$.

18. $\dfrac{7}{15} \bigcirc \dfrac{8}{13}$

19. Round 45,877,322 to the nearest ten million.

20. Give the coordinates of points E, F and G.

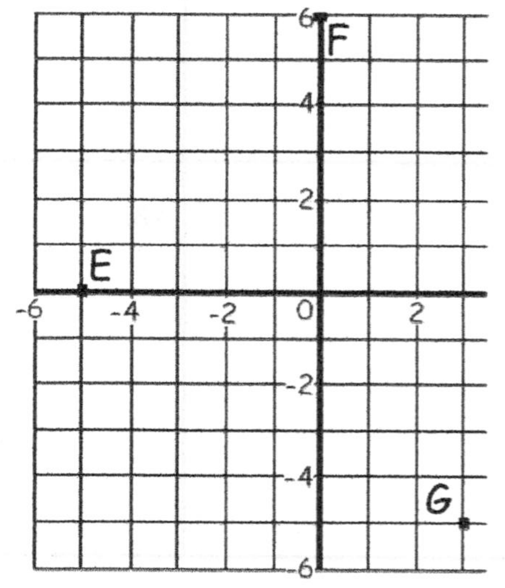

1.	2.	3.	4.
5.	6.	7.	8.
9.	10.	11.	12.
13.	14.	15.	16.
17.	18.	19.	20.

Lesson #102

1. $40 \div 5 + 6 \cdot 4 - 6 + 2 = ?$

2. Find the area of a parallelogram with a base of 17 ft. and a height of 3 ft.

3. What is the value of b? $b - 34 = 85$

4. The marbles in a bag are white, blue, or red. There are twice as many white marbles as red marbles, and there are 5 more white marbles than blue ones. There are 7 blue marbles. How many of each color are there?

5. Solve for x. $\frac{x}{6} = -1$

6. What is the P(H, H, T, T, H) on five flips of a coin?

7. Find the value of x. $5x - 9 = 36$

8. 80% of what number is 64?

9. Solve for x. $5x - 3 = 2x + 12$

10. $22 + (-16) + 8 = ?$

11. Solve for x. $\frac{x}{12} = \frac{60}{45}$

12. How many feet are in 6 yards?

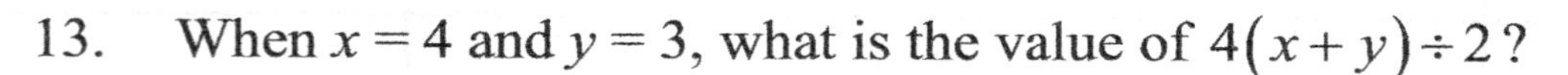

13. When $x = 4$ and $y = 3$, what is the value of $4(x + y) \div 2$?

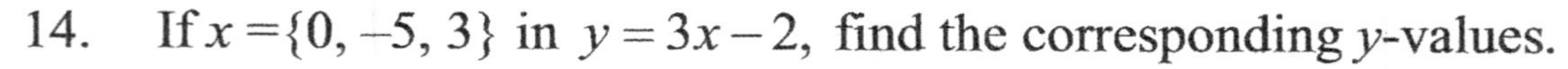

14. If $x = \{0, -5, 3\}$ in $y = 3x - 2$, find the corresponding y-values.

15. What is the value of u? $\frac{u}{7} > 5$

16. Solve for x. $7x + 4 - 15x = 36$

17. The measures of complementary angles add up to ______________.

18. Give the length of the line segment in inches.

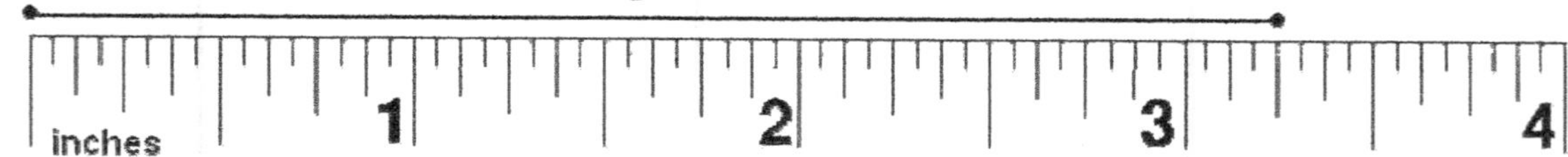

19. Find the value of x. $24 < -3x$

20. Simplify. $\frac{10x^3y^2z^2}{25xyz}$

1.	2.	3.	4.
5.	6.	7.	8.
9.	10.	11.	12.
13.	14.	15.	16.
17.	18.	19.	20.

Lesson #103

1. Solve the compound inequality. $2x + 3 < 9$ *or* $3x - 6 > 12$

2. $-75 - (-70) = ?$

3. Identify the shapes to the right. a) b)

4. Find the value of b. $b + 19 = 26$

5. Solve for x. $\frac{2}{5} = \frac{x}{65}$

6. $25 \div 5 + 7 \cdot 3 - 6 \div 2 = ?$

7. What is the value of x? $\frac{x}{9} = 13$

8. $\frac{-54}{6} = ?$

9. Find the value of $a(b+2) \div 5$ when $a = 5$ and $b = 3$.

10. $\frac{9}{10} \div \frac{3}{2} = ?$

11. A circle has a radius of 6 mm. What is the area of the circle?

12. Write 6% as a decimal and as a reduced fraction.

13. Find the missing measurement, x.

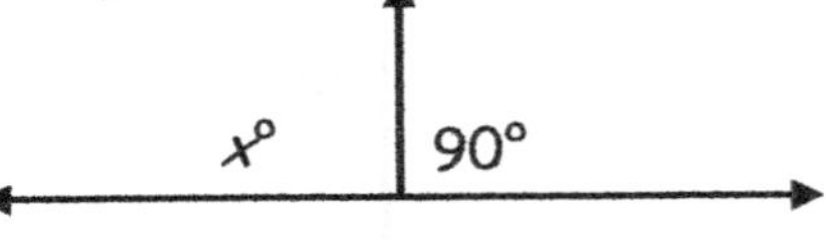

14. Find $\frac{2}{7}$ of 21.

15. $92 + (-55) = ?$

16. A baby weighs 115 ounces. What is the baby's weight in pounds and ounces?

17. Round 8.237 to the nearest hundredth.

18. Solve for a. $10a - 8 = 4a + 16$

19. $34 \times 65 = ?$

20. Find the coordinates of points A, B, and C.

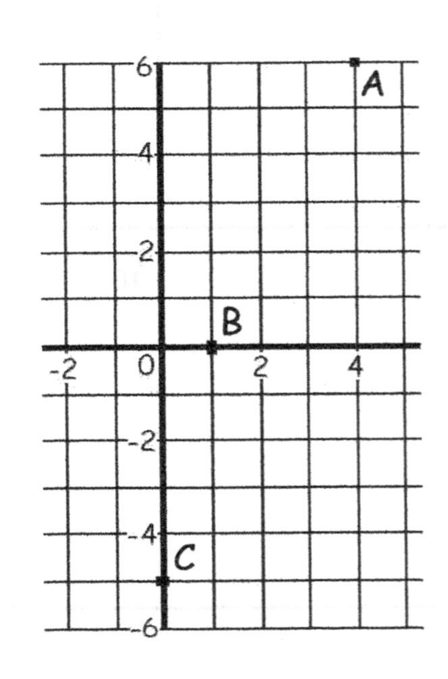

1.	2.	3.	4.
5.	6.	7.	8.
9.	10.	11.	12.
13.	14.	15.	16.
17.	18.	19.	20.

Lesson #104

1. What is the value of n? $4n+9=-19$
2. $5[4+2^2+3(4+2)]=?$
3. What are the y-values when $x=\{0, -7, 4\}$ in the equation, $y=x-7$?
4. Write an algebraic expression that means *a number decreased by five, divided by six.*
5. Solve and graph the solution on a number line. $2<-8s$
6. Solve the compound inequality. $-14<x-8<-1$
7. Use the spinner to find the P(prime number, 1)

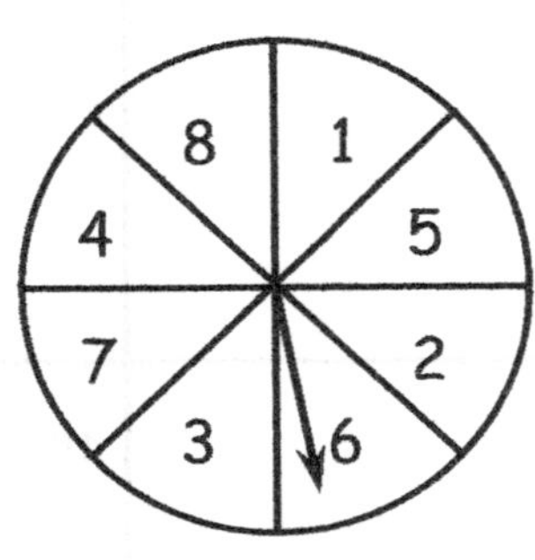

8. Find the value of x. $x+19=63$
9. $8{,}123-3{,}786=?$
10. $-|-20|=?$
11. $37-(-30)=?$

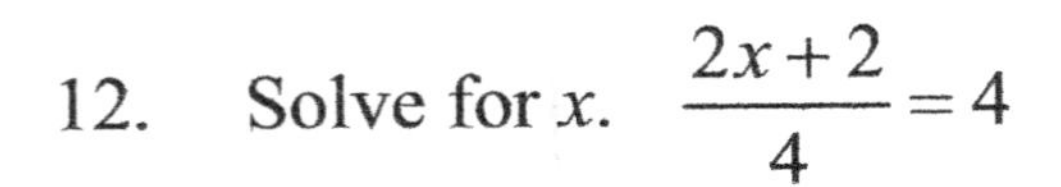

12. Solve for x. $\dfrac{2x+2}{4}=4$
13. Give the area of a square if a side measures 12 m.
14. Luke plays b-flat every fourth beat and Lindsey plays b-flat every fifth beat. The song they are playing is sixty beats long. On which beats will they both play b-flat?
15. $0.009\times 0.003=?$
16. Find the value of x. $3x-9=2x-18$
17. Find the LCM of $12a^3b^4c^2$ and $18a^2b^3c$.
18. Simplify. $7(3a+5b-6)$
19. $\dfrac{5}{9}\times\dfrac{12}{25}=?$
20. How many feet are in 4 miles?

1.	2.	3.	4.
5.	6.	7.	8.
9.	10.	11.	12.
13.	14.	15.	16.
17.	18.	19.	20.

Lesson #105

1. Find the range of 88, 19, 24 and 96.

2. $60{,}000 - 29{,}318 = ?$

3. The answer to an addition problem is the ________________.

4. Write the phrase *six times a number* using algebraic symbols.

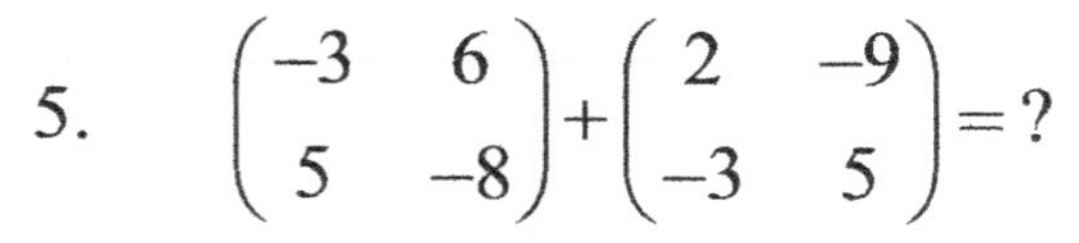

5. $\begin{pmatrix} -3 & 6 \\ 5 & -8 \end{pmatrix} + \begin{pmatrix} 2 & -9 \\ -3 & 5 \end{pmatrix} = ?$

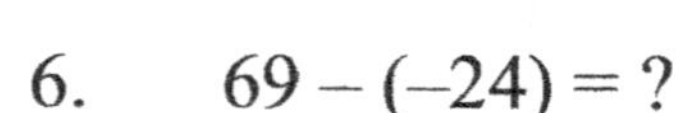

6. $69 - (-24) = ?$

7. $3 \div 0.06 = ?$

8. What is the value of x? $\frac{x}{9} + 6 = 10$

9. Solve for x. $-9x = 108$

10. Find $\frac{5}{6}$ of 24.

11. $24 + (-12) + (-10) = ?$

12. Two angles whose measures add up to 180° are ____________ angles.

13. Evaluate $3(a - b)$ when $a = 8$ and $b = 5$.

14. Use the protractor below to find the measures of ∠AOD and ∠AOF.

15. Write $\frac{9}{50}$ as a decimal and as a percent.

16. Write 42,590,000 in scientific notation.

17. If the diameter of a circle is 36 mm, what is its radius?

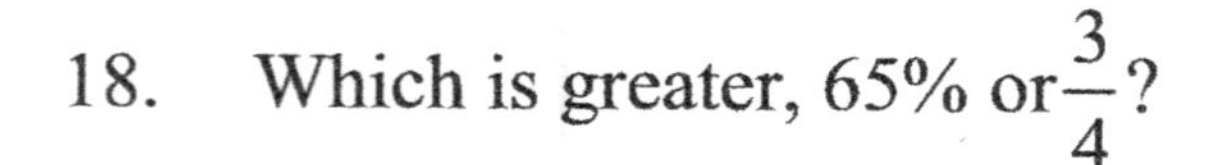

18. Which is greater, 65% or $\frac{3}{4}$?

19. $-3(9)(-2) = ?$

20. $19 - 11\frac{3}{7} = ?$

1.	2.	3.	4.
5.	6.	7.	8.
9.	10.	11.	12.
13.	14.	15.	16.
17.	18.	19.	20.

Lesson #106

1. $9 + 2[3 + 3 \cdot 4 - 2] = ?$
2. How many centimeters are in 7 meters?
3. Write the formula for calculating the area of a trapezoid.
4. Solve for x. $-5x = 75$
5. Find the missing measurement, x.

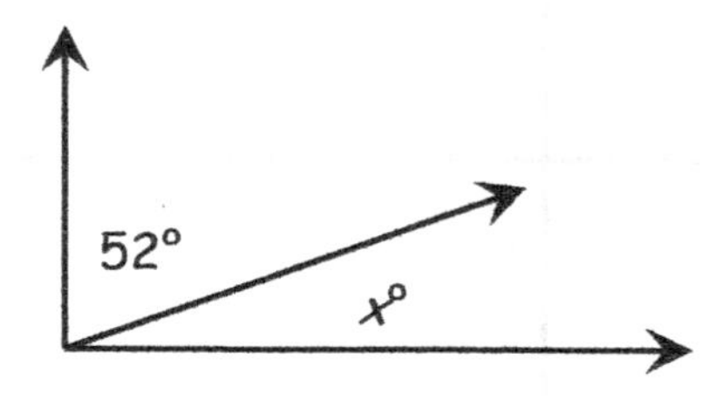

6. Solve for x. $\frac{x}{10} = 10$
7. Nine hundred years are __________ centuries.
8. Put these units in increasing order of length.

   kilometer meter millimeter

9. Subtraction Rule for Integers: Change the sign of the _________ number and ____________.
10. Write 0.0000574 in scientific notation.

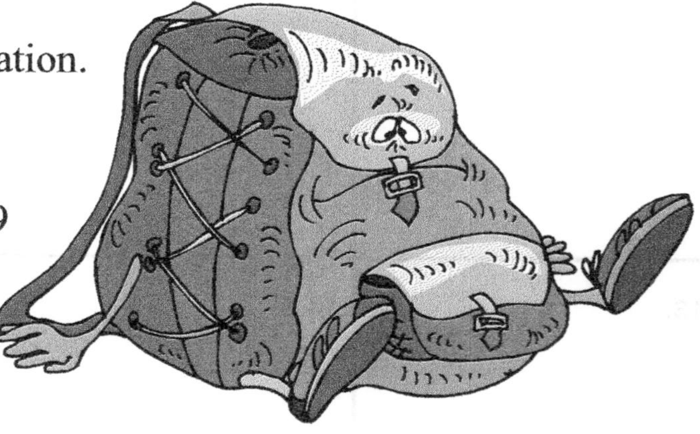

11. $90 - (-63) = ?$
12. What is the value of x? $-3x \leq -9$
13. Solve for a. $a + 49 = 100$
14. $37 + (-18) + 10 = ?$
15. Solve for x. $\frac{1}{8}x + 4 = 12$
16. $-4(9) = ?$
17. Which is greater, $\frac{6}{25}$ or 25%?
18. What is the area of a triangle if its base is 22 cm and its height is 4 cm?
19. $12\frac{2}{3} + 9\frac{2}{5} = ?$
20. Find the coordinates of points B, E, and G.

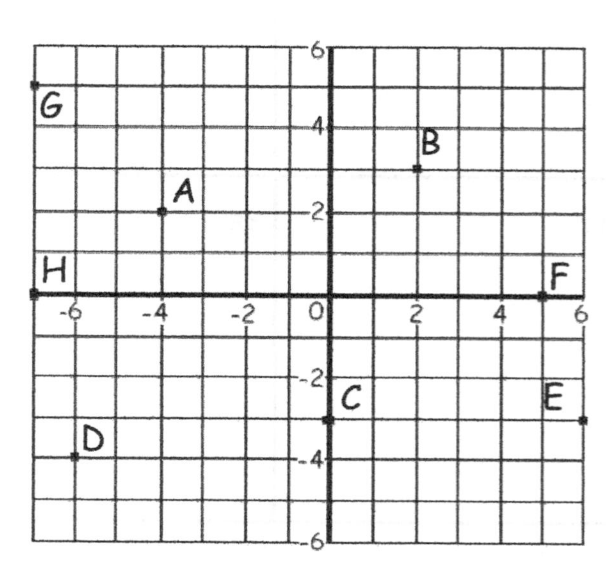

1.	2.	3.	4.
5.	6.	7.	8.
9.	10.	11.	12.
13.	14.	15.	16.
17.	18.	19.	20.

Lesson #107

1. $24 + (-12) + (-10) = ?$

2. What are the y-values in the equation, $y = x^2 + 1$, when $x = \{0, -3, 2\}$.

3. For a fund raiser, the sixth grade band members must average $60 in candy sales. Fifteen students have sold a total of $840.00. How much will the last student have to sell to raise the mean to $60 per student?

4. Graph the solution on a number line. $-5 > b - 1$

5. $15 \div 0.03 = ?$

6. Find the value of a. $\frac{1}{3}a - 6 = 15$

7. Round 382,466,019 to the nearest ten million.

8. A five-sided polygon is a(n) _____________.

9. What is the value of a? $-7a = 105$

10. Rewrite the phrase *twelve divided by a number, increased by negative five* using algebraic symbols.

11. Solve for x. $\frac{3}{15} = \frac{x}{75}$

12. A rectangular swimming pool is 4 ft deep, 15 ft wide, and 35 ft long. What is the volume of the swimming pool?

13. Find the GCF of $16x^2y^2$ and $24xy^2$.

14. Write 72,000,000 in scientific notation.

15. Solve for x. $x - 4 = 12 + 3x$

16. Find the average of 83, 87, 91 and 95.

17. What is the name for an angle that measures greater than 90°?

18. What is the P(2, 1) on two rolls of a die?

19. $97 - (-41) = ?$

20. Simplify. $\frac{4a^3b^2c^3}{8a^2bc^2}$

1.	2.	3.	4.
5.	6.	7.	8.
9.	10.	11.	12.
13.	14.	15.	16.
17.	18.	19.	20.

Lesson #108

1. Solve for x. $\frac{x}{5}=-15$

2. Give the circumference of a circle with a diameter of 8 mm.

3. Put these decimals in increasing order.

   8.2 8.02 8.0 8.32

4. $6(-3)(4)=?$

5. Write the formula for finding the surface area of a prism.

6. How many yards are in 2 miles?

7. Graph on a number line. $x-2>8$

8. What is the value of a? $a+19=25$

9. Chris' yearly college expenses were $18,000. Use the chart to find the amount Chris spent on books.

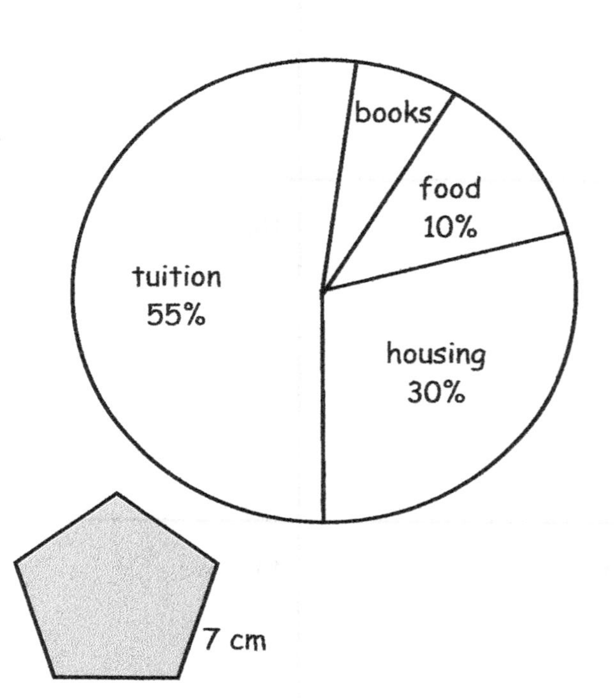

10. Find the value of x. $3x=120$

11. Solve for x. $\frac{5}{6}x-\frac{4}{6}x+7=10$

12. Find the perimeter of the pentagon.

7 cm

13. $0.7-0.3421=?$

14. Find $\frac{3}{7}$ of 35.

15. 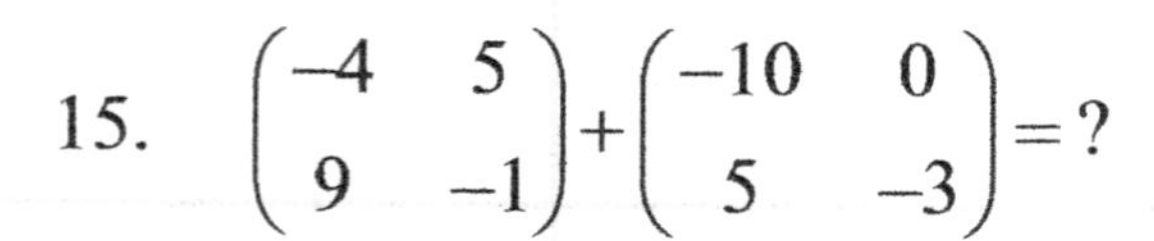

$$\begin{pmatrix}-4 & 5\\ 9 & -1\end{pmatrix}+\begin{pmatrix}-10 & 0\\ 5 & -3\end{pmatrix}=?$$

16. $3{,}475\times 6=?$

17. Find the y-values in the table.

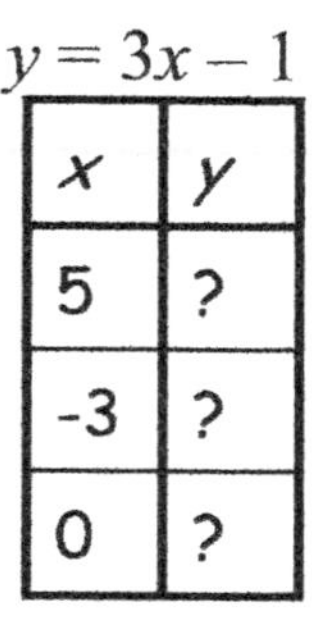

$y=3x-1$

x	y
5	?
-3	?
0	?

18. $30+4[5+4(2)]=?$

19. Write 76% as a decimal and as a reduced fraction.

20. Write the number in standard form. 7.5×10^7

1.	2.	3.	4.
5.	6.	7.	8.
9.	10.	11.	12.
13.	14.	15.	16.
17.	18.	19.	20.

Lesson #109

1. $2(w + 2) - 3w \geq 1$ Graph the solution on a number line.
2. $-66 - (-25) = ?$
3. Find the GCF of $20x^3y^2z^2$ and $25x^2yz^2$.
4. $13\frac{2}{5} - 10\frac{4}{5} = ?$
5. Solve for x. $3x - 8 = 5x + 8$
6. $-8(-4)(-2) = ?$
7. Find the value of x. $\frac{x}{7} + 14 = 21$
8. Simplify. $\frac{4a^3b^2c}{12ab^2}$
9. What is the value of x? $-5x = 75$
10. Solve for b. $b - 28 = 73$

11. Write the number in standard form. 1.685×10^{-4}
12. Find the value of y in the equation $y = 4x - 1$ when $x = \{3, 0, 2\}$.
13. What is the P(T, T, H, H) on 4 flips of a coin?
14. Solve for a. $\frac{6a + 6}{3} = -14$
15. Simplify. $8(3x - 4y + 7) - 2(3x + 2y + 8)$
16. What is 25% of 80?
17. Solve for x. $\frac{1}{7}x + 10 = 21$
18. Find the value of x. $|x| + 5 = 11$
19. $0.5 - 0.279 = ?$
20. What is the percent of change from \$12 to \$19? Round your answer to the nearest whole number.

1.	2.	3.	4.
5.	6.	7.	8.
9.	10.	11.	12.
13.	14.	15.	16.
17.	18.	19.	20.

Lesson #110

1. Solve for a. $a + 8 = 2a - 12$

2. What is the value of $5x - 2y$ when $x = 4$ and $y = 5$?

3. **The volume of a cylinder is the product of the area of a base *B* and the height *h*.** The formula is $V = B \cdot h = \pi \cdot r^2 \cdot h$. Find the volume of this cylinder with a diameter of 6 cm and a height of 9 cm.

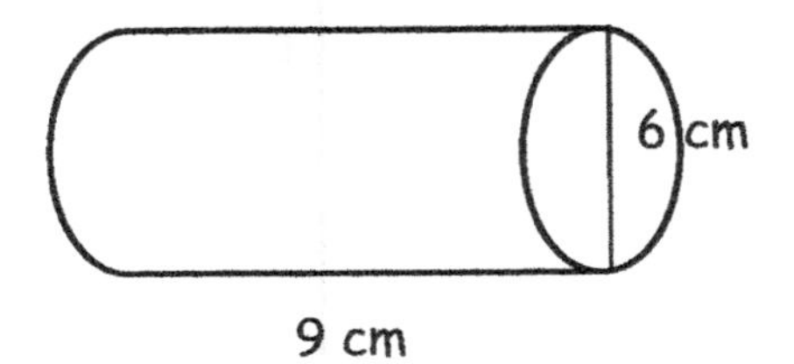

4. What is the value of x? $4x - 10 = 14$

5. 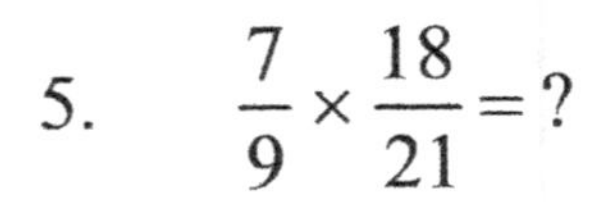 $\frac{7}{9} \times \frac{18}{21} = ?$

6. Solve for s. $2 < s - 8$

7. Write 68,000,000 in scientific notation.

8. The price of a new television is $999 and the sales tax is 7% of the price. What is the total cost of the television?

9. Write $\frac{3}{25}$ as a decimal and as a percent.

10. Find the values for y in the equation, $y = 4x + 2$ when $x = \{5, -1, 0\}$.

11. What is the value of b? $b + 29 = -75$

12. Solve for r. $|r - 8| = 5$

13. How many centimeters are in 8 meters?

14. Solve the inequality for h. $5 \leq 11 + 3h$

15. Find the area of the trapezoid.

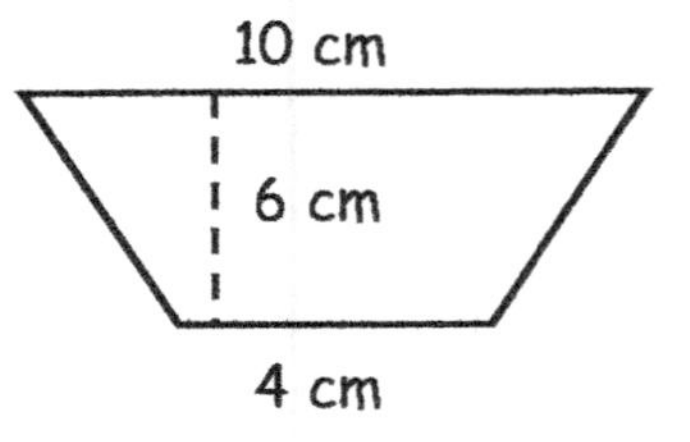

16. Write *two times a number divided by four* as an algebraic expression.

17. $32 \div 4 + 3 \cdot 5 - 4 = ?$

18. Solve for x. $\frac{1}{7}x = 21$

19. $28{,}211 - 19{,}475 = ?$

20. $5\frac{1}{2} + 3\frac{2}{5} = ?$

1.	2.	3.	4.
5.	6.	7.	8.
9.	10.	11.	12.
13.	14.	15.	16.
17.	18.	19.	20.

Lesson #111

1. Solve for n. $4n - 6 = 6n + 14$

2. $13 - 7\frac{4}{7} = ?$

3. What is the value of x? $\frac{-3x}{8} = -12$

4. Solve the proportion for x. $\frac{5}{7} = \frac{x}{105}$

5. $-4 < r - 5 \leq -1$ Graph the solution on a number line.

6. $3(-5)(2) = ?$

7. Find the LCM of $12a^3b^4c^2$ and $18a^2b^2c$.

8. $0.6 \div 0.12 = ?$

9. Write the number in standard form. 3.03×10^4

10. Simplify. $\frac{14x^2y^5z}{21xy^3z^2}$

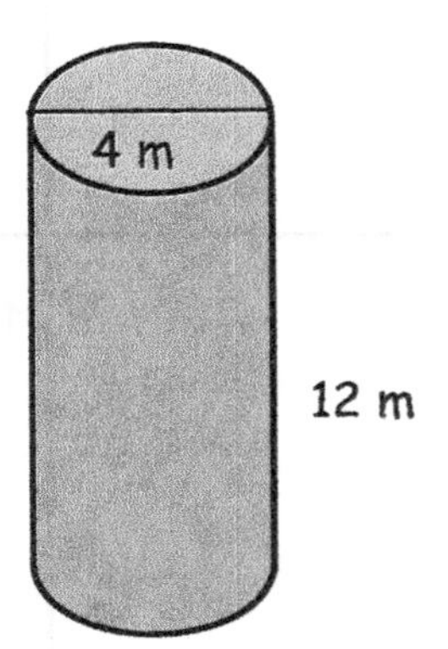

11. Find the volume of the cylinder.

12. Nikki bought a pair of earrings that cost \$15. Two weeks later they were selling for \$24. What was the percent of increase in the earrings' cost?

13. Solve for x. $x - 12 = 40$

14. $39 - (-21) = ?$

15. Solve for x. $4x + 2x - 5 = 19$

16. Find the missing measurement.

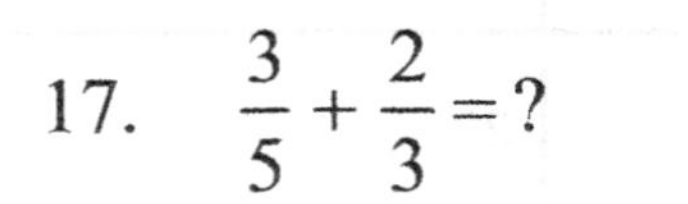

17. $\frac{3}{5} + \frac{2}{3} = ?$

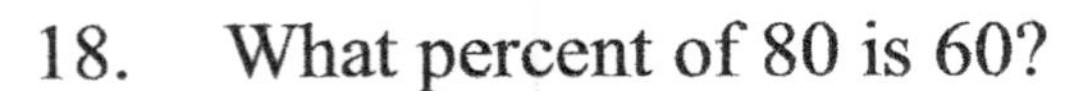

18. What percent of 80 is 60?

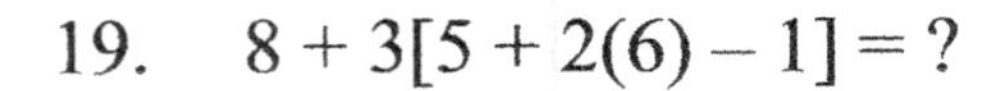

19. $8 + 3[5 + 2(6) - 1] = ?$

20. Give the coordinates for each point.

A) _____ B) _____ C) _____

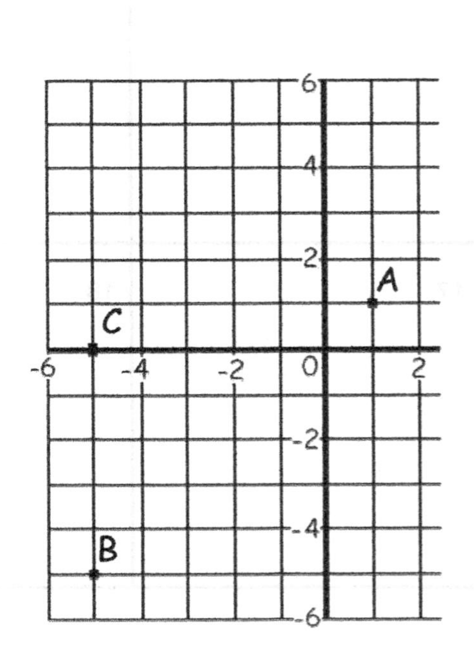

1.	2.	3.	4.
5.	6.	7.	8.
9.	10.	11.	12.
13.	14.	15.	16.
17.	18.	19.	20.

Lesson #112

1. What is the GCF of $3a^4b^2$ and $12a^2b$?

2. $86 - (-24) = ?$

3. Solve for y. $7y - 7 = 5y + 13$

4. Solve to find the value of n. $\frac{3n+2}{5} = -5$

5. $4x + 3 < -5$ Graph the solution on a number line.

6. Determine the area of this triangle.

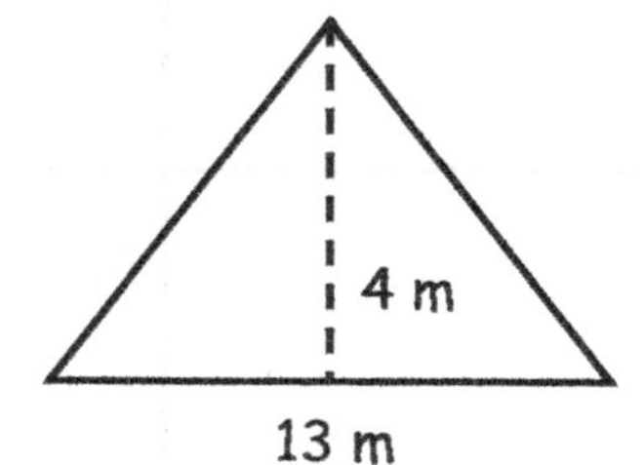

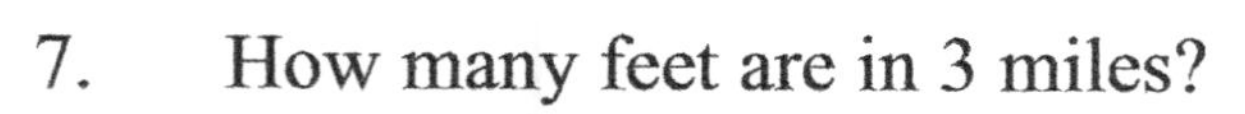

7. How many feet are in 3 miles?

8. Put these integers in order from least to greatest. −19, 13, −6, 0

9. The cost of parking in a downtown parking garage can be figured by using the formula, $C = 5 + 2(t - 1)$, where C = cost in dollars and t = # of hours in the garage. According to this formula, how much will it cost to park in the garage for 6 hours?

10. Solve for x. $|x - 3| \geq 4$

11. $\frac{-560}{5} = ?$

12. Solve for x. $-3 < x + 2 < 7$

13. 21 is what percent of 70?

14. Solve for x. $\frac{x}{9} - 12 = 20$

15. $10 - 8\frac{4}{9} = ?$

16. Find the median of 15, 86, 54, 29 and 18.

17. If $x = 6$ and $y = 10$, what is the value of $\frac{xy}{5} + 3$?

18. Find $\frac{5}{8}$ of 64.

19. What is the percent of change from 21 inches to 35 inches? Round to the nearest whole number.

20. How many pounds are 6 tons?

1.	2.	3.	4.
5.	6.	7.	8.
9.	10.	11.	12.
13.	14.	15.	16.
17.	18.	19.	20.

Lesson #113

1. Solve for x. $7x - 8 = 13$

2. What is the value of y? $\frac{1}{5}y = 22$

3. Find 85% of 40.

4. Seven decades are how many years?

5. Find the value of a. $23a + 9 = 4a + 66$

6. Solve for y. $\frac{y+4}{3} = -1$

7. $-56 + (-38) = ?$

8. $|2y - 3| \geq 7$ Graph on a number line.

9. $0.009 \times 0.03 = ?$

10. Solve for a. $5a + 9 \geq 14a - 9$

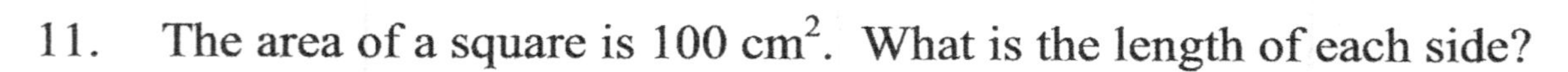

11. The area of a square is 100 cm^2. What is the length of each side?

12. Write the formula for finding the volume of a cylinder.

13. Find the LCM of $9a^4b^2c$ and $12a^2bc^2$.

14. Simplify. $8(3x - 4y) + 2(4x - 7)$

15. Solve for x. $\frac{x}{7} - 5 = 15$

16. $33 - (-17) = ?$

17. $60 + 4 \cdot 4 - 10 \div 2 + 1 = ?$

18. Write 0.36 as a percent and as a reduced fraction.

19. Write 0.00000076 in scientific notation.

20. What are the y-values in the equation, $y = x + 5$ when $x = \{0, -2, 5\}$?

1.	2.	3.	4.
5.	6.	7.	8.
9.	10.	11.	12.
13.	14.	15.	16.
17.	18.	19.	20.

Lesson #114

1. Find the circumference of a circle whose diameter is 12 mm.

2. Solve the inequality for x. $8 + x > 4$

3. Write $\frac{3}{5}$ as a decimal and as a percent.

4. $-47 - (-11) = ?$

5. $3m > 5m + 12$ Graph on a number line.

6. Of the 248 students who attended the dance, 75% were female. How many students at the dance were female? How many were male?

7. What is the value of x? $9x - 7x + 4 = 12$

8. Solve to find the value of b. $b - 29 = -86$

9. $3\frac{1}{2} + 5\frac{1}{8} = ?$

10. Simplify. $\frac{8m^3n^2p}{12m^2n^2p^2}$

11. Solve for a. $|a - 5| = 12$

12. Water freezes at ______°C.

13. What is the percent of change from \$24 to \$36?

14. Write an algebraic equation for *three times a number plus seventeen is thirty-two.*

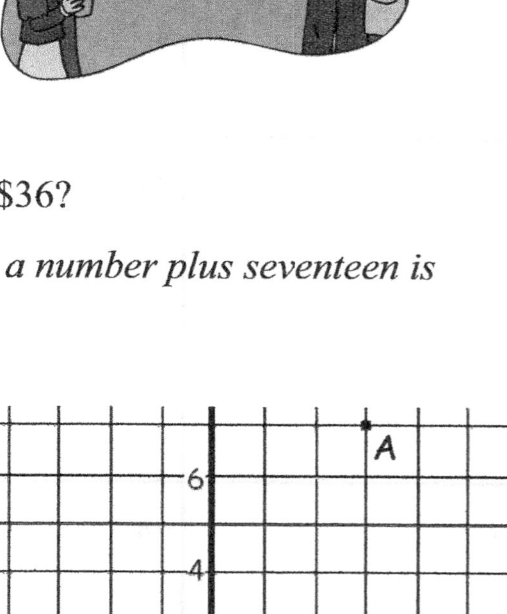

15. Solve for x. $\frac{11x - 3}{6} = 5$

16. $30 \div 6 + 4 \cdot 3 - 5 = ?$

17. Solve for x. $4x - 7 = 2x + 7$

18. $-56 + (-24) = ?$

19. Solve for x. $-6x = 72$

20. Find the coordinates of:

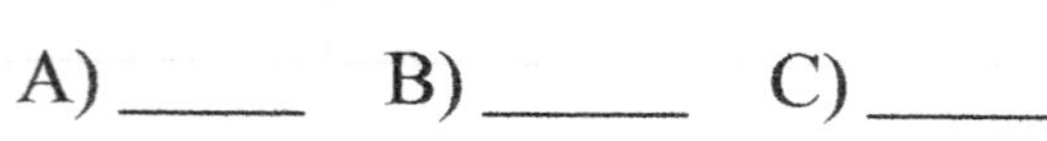

A) _____ B) _____ C) _____

1.	2.	3.	4.
5.	6.	7.	8.
9.	10.	11.	12.
13.	14.	15.	16.
17.	18.	19.	20.

Lesson #115

1. Solve to find x. $3x + 15 = 24$

2. $w + 2 > -1$ Graph the solution on a number line.

3. Rewrite *eight times a number increased by twelve* using algebraic symbols.

4. Factor. $15x^2y$.

5. What is the value of x? $-2x = 24$

6. Solve for x. $\frac{x}{4} + 15 = 25$

7. $-62 \bigcirc -35$

8. $87 + (-52) = ?$

9. Find the GCF of $15x^4y^5z^2$ and $20x^3y^2z$.

10. James got 15 out of 25 problems correct on his social studies test. What percent of the problems did he get correct?

11. Water boils at _______°F.

12. Solve for x. $-7x - 8 > 5x + 16$

13. $8\frac{3}{5} + 9\frac{1}{4} = ?$

14. Find the value of x. $17 + x = 7x - 13$

15. Which is greater, $\frac{1}{4}$ or 0.35?

16. $2.412 \div 0.04 = ?$

17. Solve for x. $3x + 2x - 5 = 10$

18. Find the value of x. $\frac{8}{9}x - \frac{7}{9}x + 3 = 12$

19. Write 6.1×10^{-3} in standard form.

20. What is the value of y in the equation, $y = 5x - 3$ when $x = \{0, -2, 1\}$?

WORLD MAP

1.	2.	3.	4.
5.	6.	7.	8.
9.	10.	11.	12.
13.	14.	15.	16.
17.	18.	19.	20.

Lesson #116

1. $2 < s - 8$ Solve the inequality for s.

2. Simplify. $8(3x - 4y - 7) + 3(2x - 2y - 5)$

3. Solve for b. $9b + 6 = 6b - 15$

4. Find the value of x. $\frac{x}{4} = -12$

5. $7(3)(-5) = ?$

6. How many meters are 700 centimeters?

7. Solve for x. $\frac{4}{5}x = 20$

8. $8 - 3\frac{5}{8} = ?$

9. Evaluate $x(y + 4)$ when $x = 5$ and $y = 4$.

10. Simplify. $\frac{18a^3b^4}{24a^2b^2}$

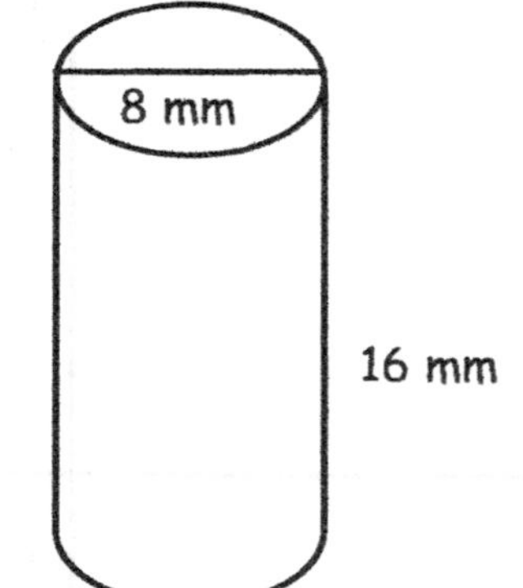

11. $92 + (-76) = ?$

12. What is the P(H, T, H, H, H) on 5 flips of a coin?

13. Find the volume of the cylinder.

14. What is the value of x? $5x - 7 = 18$

15. Find the surface area of a rectangular prism with a length of 9 cm, a width of 7 cm, and a height of 3 cm.

16. Write 0.0000089 in scientific notation.

17. Determine the perimeter of a hexagon whose sides each measure 8 m.

18. Write $\frac{5}{8}$ as a decimal.

19. Solve for x. $3(2x + 4) = 24$

20. At the reptile house, the ratio of tarantulas to lizards was 7 to 9. If there were 117 lizards, how many tarantulas were there?

1.	2.	3.	4.
5.	6.	7.	8.
9.	10.	11.	12.
13.	14.	15.	16.
17.	18.	19.	20.

Lesson #117

1. Write an algebraic expression that means *the sum of a number and ten.*

2. Two angles that add up to 90° are called __________ angles.

3. 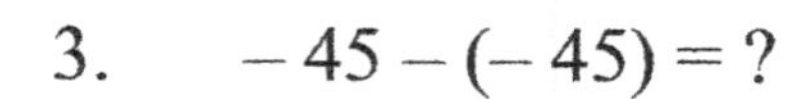
$-45-(-45)=?$

4. Simplify. $5(3m-2n-6)-3(2m+4)$

5. Solve for a. $5a+9 \geq 14a-9$

6. Find the value of b. $b+17=28$

7. Solve for x. $4x+10=2x-22$

8. Write 2.6×10^3 in standard form.

9. What is the value of x? $\frac{x}{7}+3=12$

10. 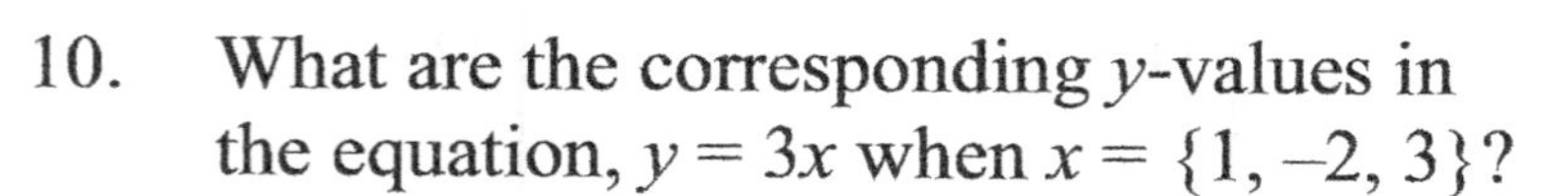
What are the corresponding y-values in the equation, $y=3x$ when $x=\{1,-2,3\}$?

11. Find the percent of change from 16 feet to 24 feet.

12. The Moran family traveled 468 miles on 18 gallons of gas. How many miles per gallon did the car average?

13. Solve for a. $7a-3=25$

14. Calculate the area of the rectangle.

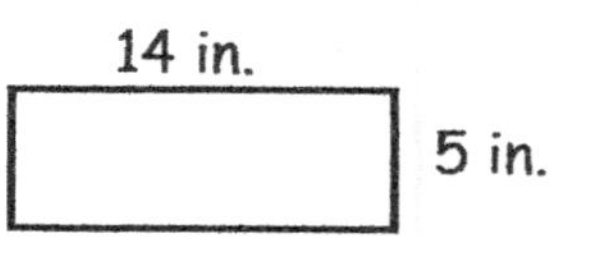

15. $\frac{5}{8} \times \frac{16}{25}=?$

16. Solve for x. $-5x=45$

17. Find the LCM of $14x^2y^2z^2$ and $16xy^2z$.

18. How many pounds are 80 ounces?

19. $\frac{-490}{-7}=?$

20. Give the coordinates for each point.

H) _____ I) ______ J) ______

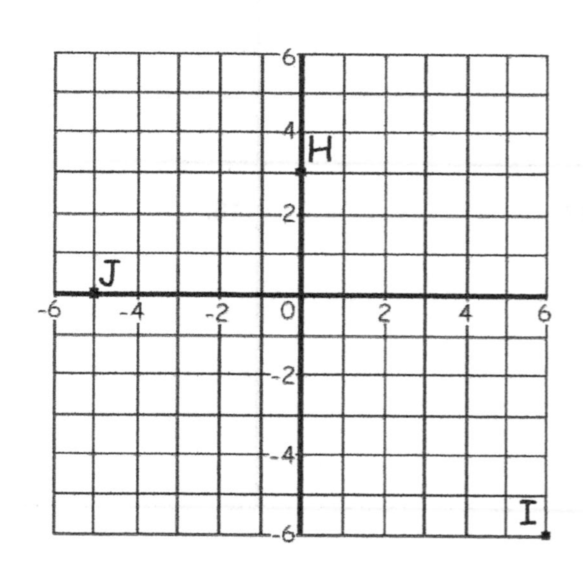

1.	2.	3.	4.
5.	6.	7.	8.
9.	10.	11.	12.
13.	14.	15.	16.
17.	18.	19.	20.

Lesson #118

1. What is the value of $6x - 2y + 8$ when $x = 3$ and $y = 2$?

2. Write 20% as a decimal and as a reduced fraction.

3. $30 \div 5 + 4 \cdot 3 - 10 \div 5 = ?$

4. Find the missing angle measurement, x.

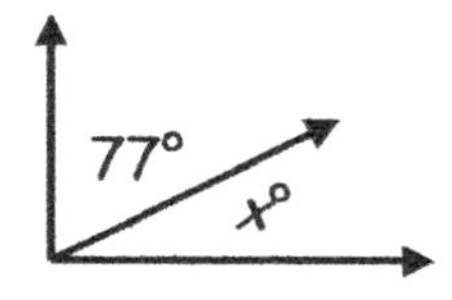

5. Solve the inequality for y. $3y - 15 > 6$

6. $-66 - (+13) = ?$

7. Solve for x. $|x| + 4 = 12$

8. Solve for y. $13y - 26 = 7y + 22$

9. A circle with a radius of 12 mm has what area?

10. Find the percent of change from 75 to 60.

11. Write 0.00236 in scientific notation.

12. Find the value of x. $\frac{1}{7}x - 12 = 15$

13. What number is 80% of 30?

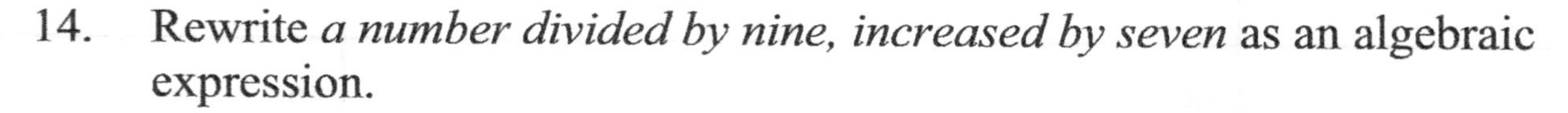

14. Rewrite *a number divided by nine, increased by seven* as an algebraic expression.

15. Becky was monitoring an experiment that involved temperature. At the beginning the temperature was 24°F. After 10 minutes, it had dropped by 21 degrees. It then rose 3° and finally dropped 14° before stabilizing, so Becky could take her final reading. What was the final reading?

16. Find the value of a. $a - 12 = 50$

17. Write the formula for finding the area of a triangle.

18. Put these decimals in increasing order. 0.5 0.05 0.42 0.2

19. $42\frac{1}{7} - 36\frac{6}{7} = ?$

20. $\begin{pmatrix} 4 & -2 & -3 \\ 7 & 6 & -2 \end{pmatrix} + \begin{pmatrix} 0 & 1 & -9 \\ 4 & -9 & 2 \end{pmatrix} = ?$

1.	2.	3.	4.
5.	6.	7.	8.
9.	10.	11.	12.
13.	14.	15.	16.
17.	18.	19.	20.

Lesson #119

1. Solve for x. $\frac{x}{5} = 14$

2. What is the value of x? $x + 8 = 2x - 12$

3. Write an algebraic phrase for *the quotient of a number and six.*

4. $\frac{4}{5} \times 1\frac{2}{3} \times 1\frac{1}{8} = ?$

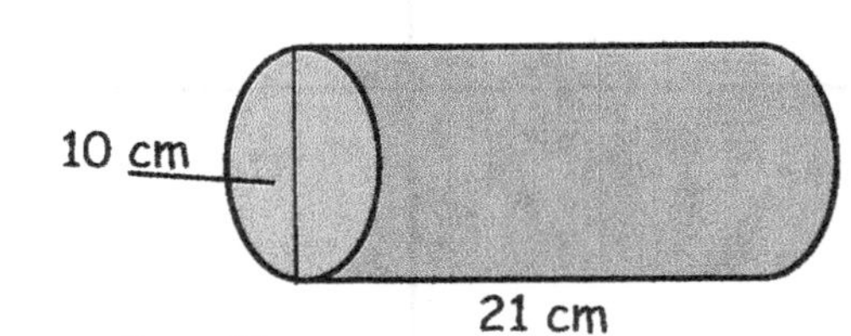

5. Find the volume of this cylinder.

6. $5 - 2x \leq 3 - x$ Graph the solution on a number line.

7. Solve for x. $8x - 5x + 4 = 40$

8. What is the value of a? $7a - 9 = 19$

9. Find the value of x. $\frac{1}{3}x = 14$

10. Find the value of 8^3.

11. Simplify. $\frac{2x^3y^2z^4}{4x^2yz^5}$

12. $1.4 \times 2.5 = ?$

13. Solve for a. $a - 25 = 100$

14. What percent of 50 is 30?

15. $37 + (-15) + (-12) = ?$

16. Simplify. $7(4a - 3b + 9)$

17. $-3(4)(5) = ?$

18. Find the circumference of a circle if its diameter is 16 cm.

19. Write 28% as a decimal and as a reduced fraction.

20. The distance from the school to the recycling center is about 825.3 yards. How many feet is it from the school to the recycling center?

1.	2.	3.	4.
5.	6.	7.	8.
9.	10.	11.	12.
13.	14.	15.	16.
17.	18.	19.	20.

Lesson #120

1. Find the slope of a line passing through points (–8, 0) and (1, 5).

2. Solve for x. $9x - 7 = 20$

3. $-88 - (-42) = ?$

4. What is the value of x? $-4x + 12 = x + 2$

5. What is the P(1, 3, 4) on 3 rolls of a die?

6. 80% of 60 is what number?

7. Find the value of x. $\frac{1}{6}x - 12 = 20$

8. At the beginning of May there were 15 sports cars on a car lot. By the end of the month, there were only 6 sports cars left on the lot. What was the percent of decrease in the number of sports cars on the lot?

9. $17\frac{1}{8} - 12\frac{7}{8} = ?$

10. Solve for x. $\frac{1}{4}x = -12$

11. How many cups are in 4 pints?

12. $130 + (-88) = ?$

13. Find the missing measurement, x.

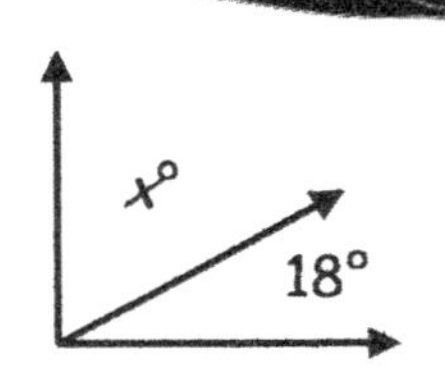

14. $20 - 2[3 + 10 \div 2] = ?$

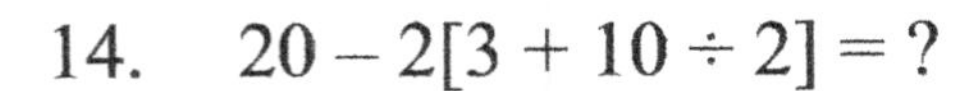

15. Calculate the area of a parallelogram if its base is 19 cm long and it is 7 cm high.

16. A triangle with sides of 5 cm, 5 cm, and 3 cm is a(n) _________ triangle.

17. $2x - 2 > 4$ Graph the solution on a number line.

18. Write 2.3×10^{-5} in standard form.

19. Write an algebraic expression for *twelve increased by six times a number.*

20. What value of x makes the two fractions equivalent? $\frac{x}{40} = \frac{80}{100}$

1.	2.	3.	4.
5.	6.	7.	8.
9.	10.	11.	12.
13.	14.	15.	16.
17.	18.	19.	20.

Lesson #121

1. Solve for x. $-8x = 27 + x$

2. Jennifer bought a sweater for \$12. This was $\frac{3}{4}$ of the regular price. What was the regular price of the sweater?

3. What is the value of x? $\frac{x}{5} - 14 = 25$

4. $60 - 5 \cdot 5 + 12 \div 4 - 1 = ?$

5. Find the slope of a line passing through points $(-4, -5)$ and $(-9, 1)$.

6. $\frac{5}{6} \bigcirc \frac{7}{8}$

7. $67 - (-23) = ?$

8. Solve the inequality for x. $7 - x \leq 12$

9. Write 0.00406 in scientific notation.

10. Use the chart to find the y-values in the equation.

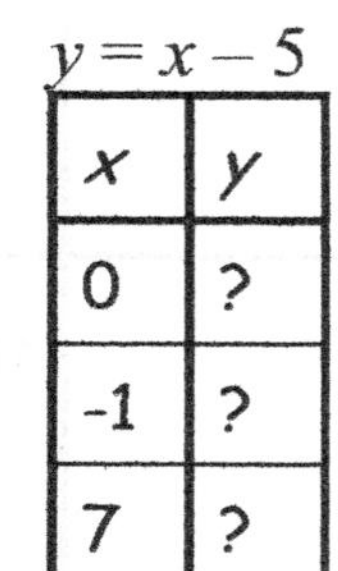

$y = x - 5$

x	y
0	?
-1	?
7	?

11. Find the area of a rectangle if its length is 14 cm and its width is 5 cm.

12. Evaluate $5y + x$ when $x = 4$ and $y = 3$.

13. $0.009 \times 0.2 = ?$

14. $\begin{pmatrix} 5 & -3 \\ 4 & 10 \end{pmatrix} - \begin{pmatrix} -9 & 7 \\ 8 & -6 \end{pmatrix} = ?$

15. How many centuries are 600 years?

16. Round 374,866,219 to the nearest million.

17. Solve for b. $b - 26 = 55$

18. Find the value of x. $5x - 12 = 18$

19. Which is greater, 55% or $\frac{3}{4}$?

20. Solve for x. $\frac{1}{8}x + 14 = 20$

1.	2.	3.	4.
5.	6.	7.	8.
9.	10.	11.	12.
13.	14.	15.	16.
17.	18.	19.	20.

Worksheet #122

1. Find the GCF of $4a^5b^2c$ and $12a^3bc^2$.

2. What percent of 50 is 10?

3. $7\frac{2}{5} + 8\frac{1}{3} = ?$

4. Solve for a. $a + 8 = 2a - 12$

5. Find the slope of the line passing through points $(-3, 7)$ and $(-1, 1)$.

6. How many millimeters are in 6 meters?

7. Find the value of x. $-6x = 72$

8. What is the value of x? $\frac{x}{7} = -13$

9. $37 - (-19) = ?$

10. $5(4x - 5y + 5) - 3(2x - 6) = ?$

11. The slope of a horizontal line is ________.

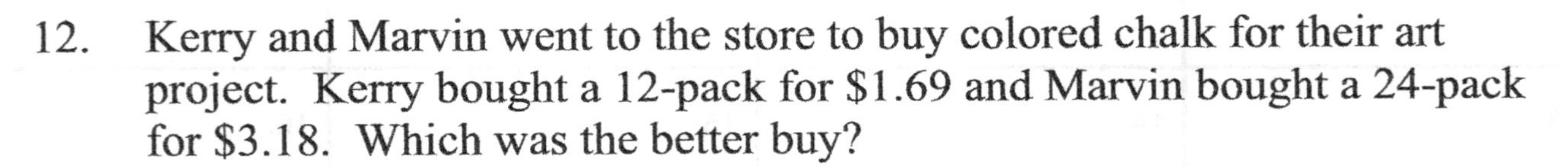

12. Kerry and Marvin went to the store to buy colored chalk for their art project. Kerry bought a 12-pack for \$1.69 and Marvin bought a 24-pack for \$3.18. Which was the better buy?

13. $4 + x > 3$ or $6x < -30$ Graph the solution on a number line.

14. Solve for x. $3x - 1 = 8$

15. $0.6 - 0.374 = ?$

16. Find the area of this triangle.

5 mm

12 mm

17. What is the value of a? $4a - a + 6 = -6$

18. Write 3.709×10^6 in standard form.

19. Write 65% as a decimal and as a reduced fraction.

20. Solve for a. $\frac{5a - 2}{8} = -4$

1.	2.	3.	4.
5.	6.	7.	8.
9.	10.	11.	12.
13.	14.	15.	16.
17.	18.	19.	20.

Lesson #123

1. Solve for x. $x + 5(x - 1) = 13$
2. Solve the inequality for x. $3x + 5 < -4$
3. Find the value of x. $-x - 3 = 1 - 3x$
4. $\frac{8}{9} \times \frac{18}{24} = ?$
5. Solve for x. $|x+8| \geq 5$
6. $50 - 4[3(2) + 6] = ?$
7. If $a = 6$ and $b = 10$, what is the value of $\frac{ab}{5} + 9$?
8. Write 2,427,000 in scientific notation.
9. Solve for x. $\frac{1}{3}x = 13$
10. $70{,}000 - 18{,}473 = ?$
11. Find the value of a. $a + 25 = -70$
12. $16 + (-12) + 18 = ?$
13. How many quarts are in 8 gallons?
14. $\frac{-8{,}420}{2} = ?$
15. What is the P(H, H, T) on 3 flips of a coin?
16. The average weight of three pigs is 248 lb. One pig weighs 250 lb. and another weighs 242 lb. How much does the third pig weigh?
17. Solve for x. $\frac{x}{9} - 8 = -15$
18. Find $\frac{4}{7}$ of 63.
19. $14\frac{1}{10} - 6\frac{4}{5} = ?$
20. Simplify. $\frac{20x^3y^2z^5}{25x^2yz^2}$

1.	2.	3.	4.
5.	6.	7.	8.
9.	10.	11.	12.
13.	14.	15.	16.
17.	18.	19.	20.

Lesson #124

1. $-9m \geq 36$ Graph the solution on a number line.

2. $13 - 9\frac{4}{5} = ?$

3. Solve for the value of w. $9w + 6 = 6w - 15$

4. Round 12.6532 to the nearest tenth.

5. Two angles whose measures add up to 180° are _________ angles.

6. Solve for x. $x + 9 = 35$

7. Find the area of this trapezoid.

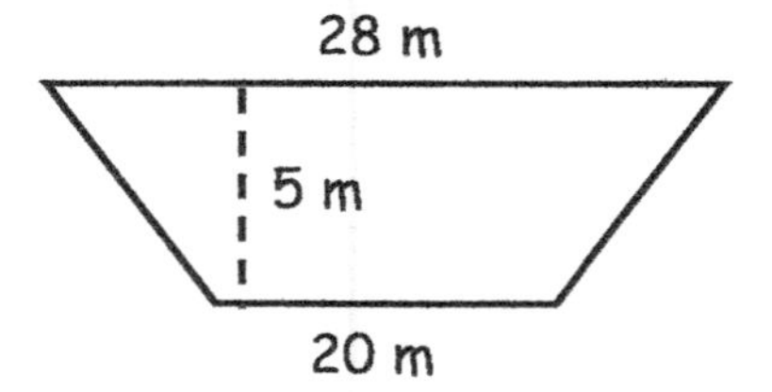

8. Find the value of y. $4(y - 8) = -12$

9. Write an expression to represent *twelve divided by two times a number.*

10. $8(5 - 2) + 12 \div 3 - 2 = ?$

11. $-37 + (-66) = ?$

12. $|g - 3| > 2$ Solve the inequality for g.

13. Use the table to find the corresponding y-values.

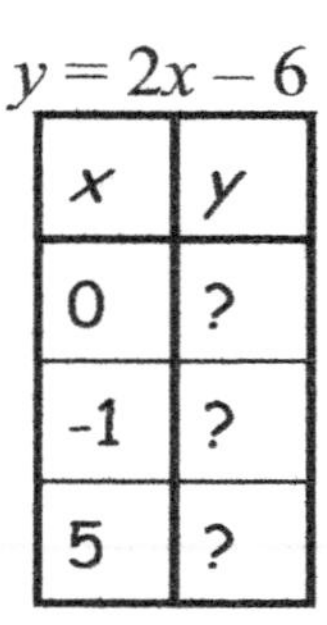

$y = 2x - 6$

x	y
0	?
-1	?
5	?

14. How many yards are in 3 miles?

15. $\begin{pmatrix} 9 & 4 & -7 \\ 8 & -2 & 9 \end{pmatrix} + \begin{pmatrix} -2 & 3 & -1 \\ 0 & -6 & 5 \end{pmatrix} = ?$

16. Solve for x. $\frac{x}{15} = 10$

17. $1.35 \times 0.5 = ?$

18. Find the value of h. $10h - 4 = -94$

19. Solve the proportion for x. $\frac{3}{7} = \frac{x}{105}$

20. Find the percent of change from \$12 to \$9.

1.	2.	3.	4.
5.	6.	7.	8.
9.	10.	11.	12.
13.	14.	15.	16.
17.	18.	19.	20.

Lesson #125

1. The slope of a horizontal line is always ____________.

2. $6\frac{2}{3} + 9\frac{2}{5} = ?$

3. Find the slope of the line through points (–2, 1) and (5, 7).

4. Solve for the value of w. $9w + 9 = 3w - 15$

5. How many pounds are 5 tons?

6. Translate into algebraic symbols: *One third of a number increased by seven.*

7. According to the Fahrenheit scale, water freezes at what temperature?

8. Find the values for y in the equation, $3x + y = 3$ when $x = \{-3, 1, 2\}$.

9. $68 - (-24) = ?$

10. Solve for x. $\frac{x}{8} - 12 = -30$

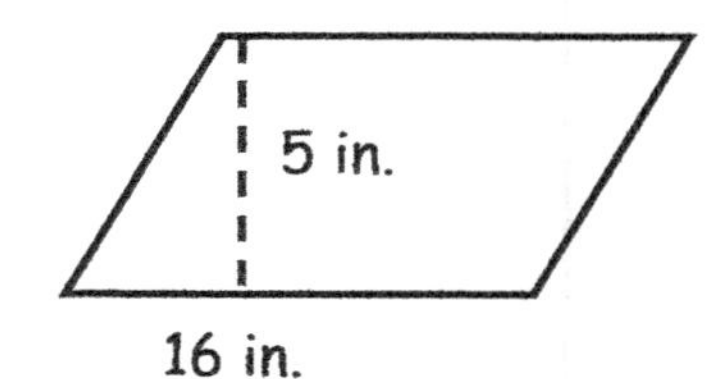

11. Find the area of the parallelogram.

12. Solve for y. $8y - 15 = 25$

13. Round 376,255,318 to the nearest hundred million.

14. $375.9 + 6.462 = ?$

15. The ratio of ducks to fish in a pond was 2 to 13. If there were 143 fish, how many ducks were in the pond?

16. Three-fourths of the 48 classrooms at Waterson Middle School have DVD players. Two-thirds of these DVD players do not work. How many DVD players work?

17. Write 8.52×10^{-4} in standard form.

18. $-63 + (-69) = ?$

19. Find the value of x. $\frac{2}{5}x = 20$

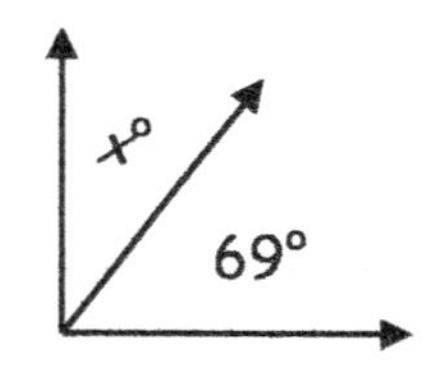

20. Find the missing measurement, x.

1.	2.	3.	4.
5.	6.	7.	8.
9.	10.	11.	12.
13.	14.	15.	16.
17.	18.	19.	20.

Lesson #126

1. *Four times a number is 52.* Write an equation to represent the sentence.

2. Find the GCF of $16a^3b^2$ and $24a^2b$.

3. Solve for x. $3x + 12 = 6x - 3$

4. What is the value of y? $y + 13 = 98$

5. Simplify. $\dfrac{20x^4y^5z^3}{25x^3y^2z}$

6. How many inches are in 5 feet?

7. Find the slope of the line passing through points (8, 3) and (– 4, 3).

8. $-9(7) = ?$

9. Last summer the price of a dress was \$125. This summer the same dress is selling for \$160. By what percent did the price of the dress increase?

10. Solve for c. $-8 < 2c + 10 < 14$

11. Find the median of 16, 49, 10, 86 and 31.

12. What is the value of x? $\dfrac{1}{5}x = -12$

13. Find the average of 65, 25 and 75.

14. Write the formula for finding the volume of a cylinder.

15. What is the P(H, T, H) on 3 flips of a coin?

16. Evaluate $9x - y$ when $x = 3$ and $y = 7$.

17. $0.008 \times 0.7 = ?$

18. $40 \div 8 + 3 \cdot 5 - 9 \div 3 = ?$

19. $\dfrac{5}{6} \times \dfrac{12}{15} = ?$

20. Write 25% as a decimal and as a reduced fraction.

1.	2.	3.	4.
5.	6.	7.	8.
9.	10.	11.	12.
13.	14.	15.	16.
17.	18.	19.	20.

Lesson #127

1. $\dfrac{-720}{3} = ?$

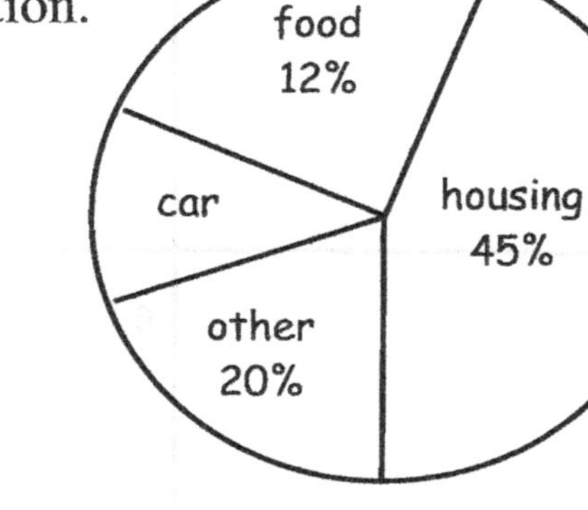

2. Write the slope-intercept form of a linear equation.

3. Round 32.77 to the nearest tenth.

4. Use the pie chart to find what percent of his income that Greg spent on his car. If Greg's income was $20,000, how much did he spend on housing?

5. $-32 + (-16) + 24 = ?$

6. Find the slope of the line passing through points (9, –2) and (3, 4).

7. Write an algebraic expression for *negative seven times a number, divided by nine.*

8. Solve for y. $4(y - 8) = -12$

9. Write $\dfrac{2}{50}$ as a decimal and as a percent.

10. $5{,}124 - 1{,}786 = ?$

11. Find the value of b. $b - 17 = -39$

12. Solve for x. $\dfrac{x}{14} = 5$

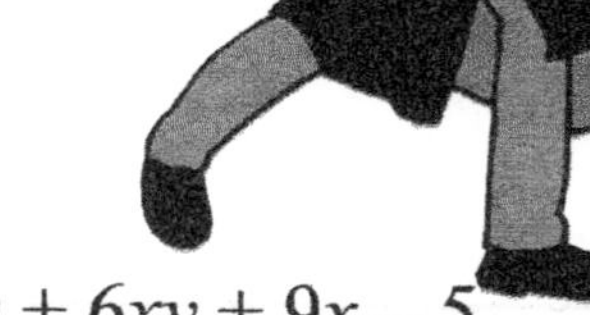

13. $\dfrac{3}{5} \times \dfrac{5}{18} = ?$

14. What is the value of y? $\dfrac{3y}{4} = 12$

15. Combine like terms. $3x + 4xy + 9 + 6xy + 9x - 5$

16. Find the area of a circle whose radius is 6 millimeters.

17. $17 - 8\dfrac{2}{7} = ?$

18. Write 6.411×10^5 in standard form.

19. Identify the slope and the y-intercept for $y = 3x - 5$.

20. Two angles whose measures add up to 90° are ____________ angles.

1.	2.	3.	4.
5.	6.	7.	8.
9.	10.	11.	12.
13.	14.	15.	16.
17.	18.	19.	20.

Lesson #128

1. What is the value of x? $\frac{1}{3}x = 21$

2. Solve for x. $6x = 21 - x$

3. $9 > 5 - 4t$ Graph the solution on a number line.

4. Jenny wants to make a bead-bracelet that is 7 inches long. Each bead is two-fifths of an inch long. How many beads are needed for the bracelet?

5. Find the slope and the y-intercept for $y = \frac{7}{6}x + \frac{3}{4}$.

6. $3.6 - 1.79 = ?$

7. Solve to find the value of x. $\frac{x}{10} + 17 = 24$

8. How many grams are in 3 kilograms?

9. Solve for x. $4x - 19 = 17$

10. Write the slope-intercept form of a linear equation.

11. Find the area of the rectangle to the right.

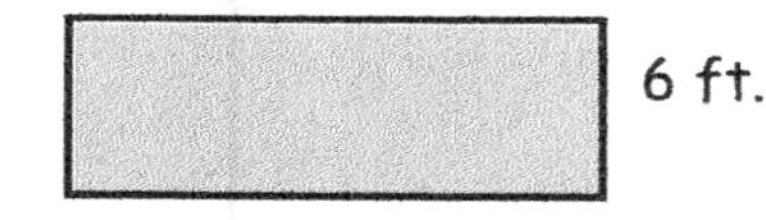

12. Evaluate $\frac{abc}{2} + 8c$ when $a = 2$, $b = 3$ and $c = 4$.

13. Find the percent of change from 40 cm to 25 cm. Round to the nearest whole number.

14. $-4 < k + 1 < 7$ Solve the inequality for k.

15. Find the slope of the line passing through points $(4, -1)$ and $(4, 7)$.

16. $37 - (-12) = ?$

17. Find $\frac{4}{5}$ of 35.

18. Solve the proportion for x. $\frac{5}{8} = \frac{x}{104}$

19. $\frac{4}{5} + \frac{1}{3} = ?$

20. Solve for y. $\frac{2y - 2}{4} = 3$

1.	2.	3.	4.
5.	6.	7.	8.
9.	10.	11.	12.
13.	14.	15.	16.
17.	18.	19.	20.

Lesson #129

1. Find the LCM of $12x^4y^5z^2$ and $14x^3y^2z^2$.

2. Solve for b. $2b + 38 = 76$

3. If $a = 4$ and $b = 3$, what is the value of $(6a - 2b) \div 2$?

4. Find the value of y. $12y + 13 = 8y + 33$

5. What is the slope of the line through points $(-2, 1)$ and $(5, 7)$?

6. Express the phrase *sixteen divided by a number and twelve more* using algebraic symbols.

7. $-56 + (-43) = ?$

8. Distribute and combine like terms. $6(3x - 4y - 5) - 3(2x + 5)$

9. Tell whether $\sqrt{2}$ is rational or irrational.

10. Find the surface area of the rectangular prism.

4 cm
2 cm
8 cm

11. Solve the inequality for b. $-5 > b - 1$

12. Write the slope-intercept form of a linear equation.

13. $\dfrac{200}{-4} = ?$

14. Solve for x. $3x = -81$

15. $3\dfrac{3}{7} + 2\dfrac{1}{2} = ?$

16. $2 \cdot 3[5 + (4 \div 2)] = ?$

17. $47 + (-25) + (-13) = ?$

18. $\dfrac{6}{8} \cdot \dfrac{12}{18} = ?$

19. 15 is what percent of 80?

20. Write 6,700,000 in scientific notation.

1.	2.	3.	4.
5.	6.	7.	8.
9.	10.	11.	12.
13.	14.	15.	16.
17.	18.	19.	20.

Lesson #130

1. Solve for x. $9x - 3x + 5 = -25$

2. The slope of a horizontal line is always ___________.

3. How many feet are in three miles?

4. Evaluate $\frac{rst}{4} - rs$ when $r = 2$, $s = 3$ and $t = 4$.

5. $-4x - 2 < 10$ Graph the inequality on a number line.

6. Solve for x. $\frac{2x}{5} = 6$

7. Find the value of x. $3x + 5 = 2x - 9$

8. Find the slope and the y-intercept for $y = 4x - 1$.

9. $-86 - (+26) = ?$

10. Write 3.5×10^{-3} in standard form.

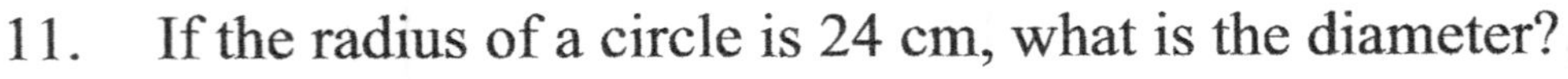

11. If the radius of a circle is 24 cm, what is the diameter?

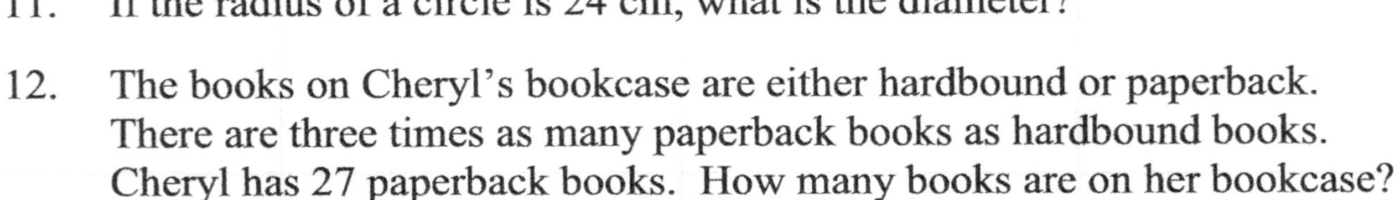

12. The books on Cheryl's bookcase are either hardbound or paperback. There are three times as many paperback books as hardbound books. Cheryl has 27 paperback books. How many books are on her bookcase?

13. $1.235 \times 0.3 = ?$

14. Two angles whose measures add up to 90° are ______ angles.

15. Simplify. $\frac{14x^3y^2z^4}{21xyz^2}$

16. Determine the slope of the line through the points (4, 2) and (–2, –1).

17. $16\frac{3}{7} - 12\frac{6}{7} = ?$

18. What number is 70% of 25?

19. $0.4 \div 0.08 = ?$

20. Write 0.05 as a percent and as a reduced fraction.

1.	2.	3.	4.
5.	6.	7.	8.
9.	10.	11.	12.
13.	14.	15.	16.
17.	18.	19.	20.

Lesson #131

1. Write the slope-intercept form for a linear equation.
2. Distribute and combine like terms. $7(3x - 4y + 5) + 6(2x - 2)$
3. Lauren was painting snowmen for the craft show. She was paid 15¢ for each snowman she painted. If Lauren was paid $9.00, how many snowmen did she paint?
4. Solve for b. $4b - 10 = 10$
5. Determine the slope and the y-intercept for $y = \frac{2}{3}x - 5$.
6. What is the value of x? $\frac{1}{9}x - 7 = 20$
7. $-137 - (-45) = ?$
8. $45 \div 5 + 8 \cdot 2 - 7 - 3 = ?$
9. Solve the inequality for b. $3b + 12 > 21 + 2b$
10. Solve for a. $-9a = 108$
11. $25\frac{1}{4} + 13\frac{2}{5} = ?$
12. Write 0.0987 in scientific notation.
13. Write $\frac{4}{5}$ as a decimal and as a percent.
14. How many ounces are in 7 pounds?
15. Which is greater, 0.02 or 5%?
16. Evaluate $\frac{2a}{b} - c$ when $a = 15$, $b = 5$ and $c = 2$.
17. Solve for x. $5x + 3 = 18$
18. Write the equation of the line through point (3, 5) when m = 2.
19. On the Celsius temperature scale, water boils at ___________.
20. For the equation, $5x + y = 7$, choose three values for x and find the corresponding values of y.

1.	2.	3.	4.
5.	6.	7.	8.
9.	10.	11.	12.
13.	14.	15.	16.
17.	18.	19.	20.

Lesson #132

1. Write an expression for *the product of a number and eight.*

2. Solve for b. $\frac{b}{5} = -5$

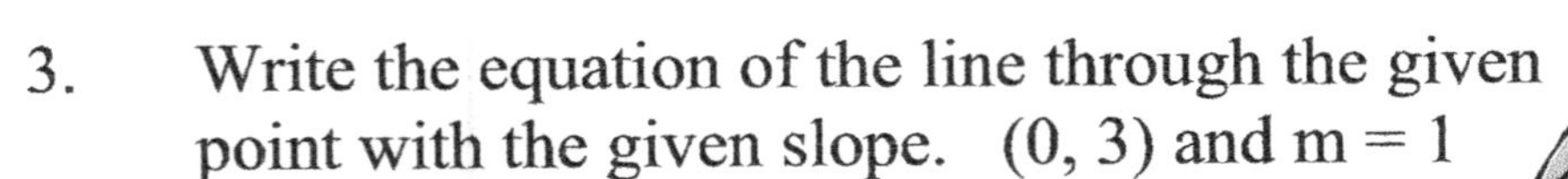

3. Write the equation of the line through the given point with the given slope. (0, 3) and m = 1

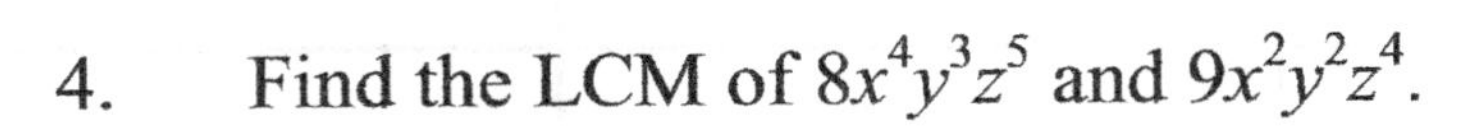

4. Find the LCM of $8x^4y^3z^5$ and $9x^2y^2z^4$.

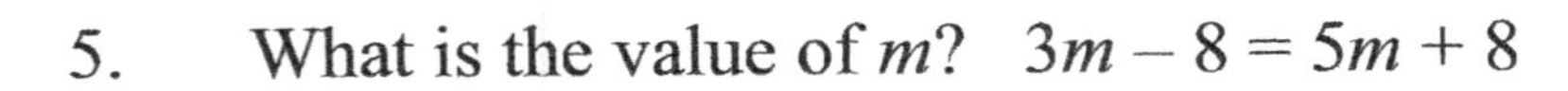

5. What is the value of m? $3m - 8 = 5m + 8$

6. $105 + (-56) = ?$

7. Solve the inequality for f. $f - 10 \leq 16$

8. What is the slope of a horizontal line?

9. Solve the proportion for x. $\frac{x}{24} = \frac{5}{6}$

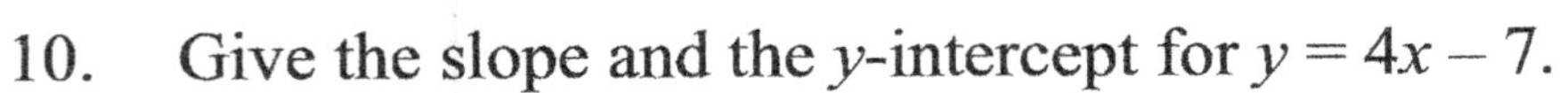

10. Give the slope and the y-intercept for $y = 4x - 7$.

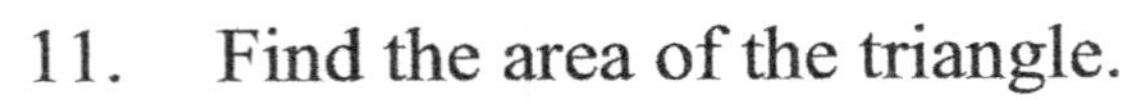

11. Find the area of the triangle.

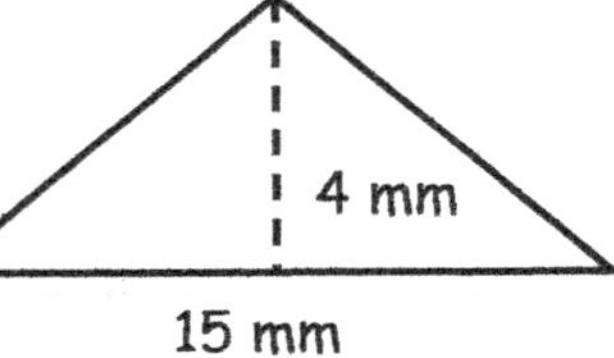

12. $-5(-6)(-2) = ?$

13. $700{,}000 - 465{,}337 = ?$

14. Find the probability of rolling a number above 2 on one roll of a die.

15. Michael began lifting weights a month ago. When he started he could lift only 165 lbs. Now he can lift 190 pounds. Find the percent of change. Round to the nearest whole number.

16. Solve for a. $a - 26 = 43$

17. The distance around the outside of a circle is the ______________.

18. Round 86.247 to the nearest hundredth.

19. A number that occurs most often in a set of numbers is the __________.

20. Find the missing measurement, x.

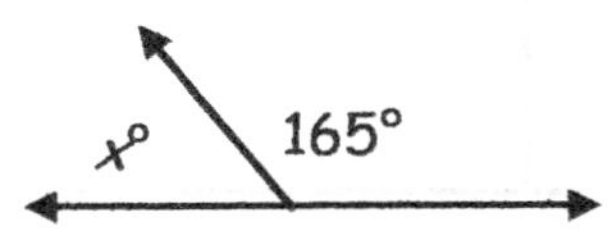

1.	2.	3.	4.
5.	6.	7.	8.
9.	10.	11.	12.
13.	14.	15.	16.
17.	18.	19.	20.

Lesson #133

1. $x - 4 \leq 2$ Graph the inequality on a number line.

2. $75 + (-30) = ?$

3. Solve for x. $3x - 10 = 11$

4. $8\frac{5}{10} - 6\frac{1}{2} = ?$

5. Solve the inequality. $-4 < x - 5 \leq -1$

6. Find the slope of the line passing through points $(-2, 0)$ and $(4, -3)$.

7. $0.6 \bigcirc 0.39$

8. Solve for b. $\frac{1}{10}b - 19 = -26$

9. How many teaspoons are in 5 tablespoons?

10. What is the probability of rolling a prime number on one roll of a die?

11. Write the equation of the line through point $(1, 2)$ if $m = -3$.

12. Find the value of x. $\frac{3}{5}x = 30$

13. Write the formula for finding the circumference of a circle.

14. Write 5.4×10^{-6} in standard form.

15. What is the slope and the y-intercept for the equation, $y = \frac{2}{3}x - 9$?

16. What number is 40% of 30?

17. Find the volume of the cylinder.

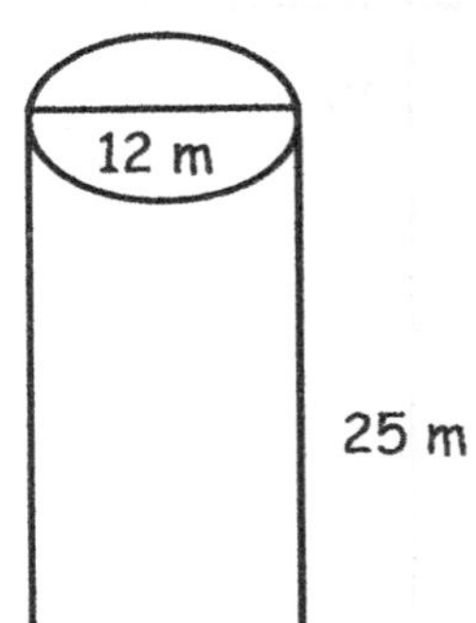

18. $16 \div 4 + 8[3 + 10 \div 5] = ?$

19. What is the value of x? $-8x = 104$

20. Solve for y. $y + 12 = -54$

1.	2.	3.	4.
5.	6.	7.	8.
9.	10.	11.	12.
13.	14.	15.	16.
17.	18.	19.	20.

Lesson #134

1. Solve the inequality for x. $-8x + 16 \leq 8$

2. Write the slope-intercept form of a linear equation.

3. $1.7 - 0.98 = ?$

4. Find the percent of change from 180 lbs. to 150 lbs. Round to the nearest whole number.

5. $4(-5)(3) = ?$

6. When $a = 5$ and $b = 3$, evaluate $3a - b + 6$.

7. Distribute and combine like terms. $7(2x - 3y + 7) - 3(3x + 4y + 2)$

8. $-150 + (-75) = ?$

9. How many decades are 80 years?

10. Simplify. $\dfrac{14x^3y^2}{21xy^2}$

11. $\begin{pmatrix} -1 & 2 & -6 \\ 7 & -3 & 8 \end{pmatrix} + \begin{pmatrix} -1 & 4 & -9 \\ 0 & -5 & 4 \end{pmatrix} = ?$

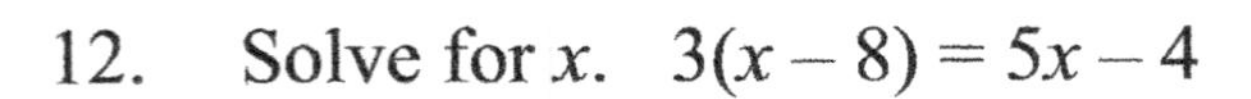

12. Solve for x. $3(x - 8) = 5x - 4$

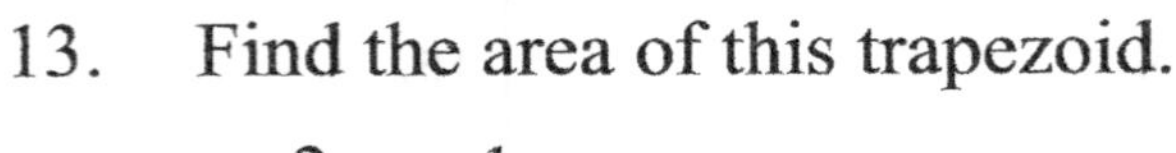

13. Find the area of this trapezoid.

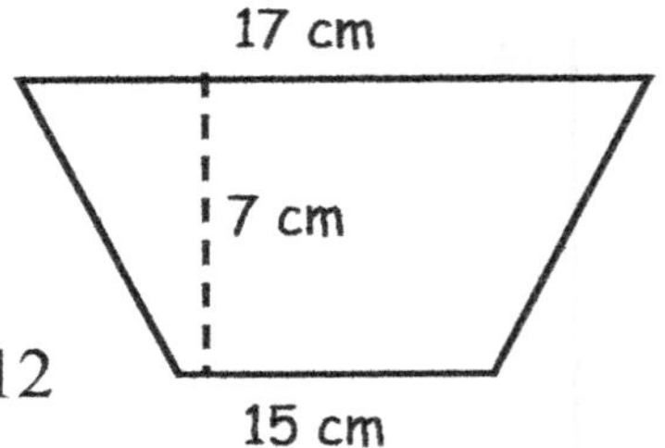

14. $13\frac{2}{5} + 8\frac{1}{4} = ?$

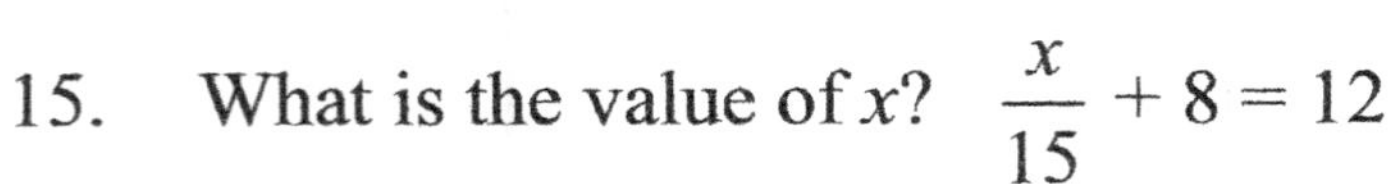

15. What is the value of x? $\dfrac{x}{15} + 8 = 12$

16. Write 92,000,000 in scientific notation.

17. Write the equation for the line through points (7, 3) and (2, 2).

18. The price of a picture frame is \$6. The total price of two picture frames and three scrapbooks is \$36. What is the price of a single scrapbook?

19. Solve for b. $b + 15 = -55$

20. $\dfrac{8}{9} \times \dfrac{18}{24} = ?$

1.	2.	3.	4.
5.	6.	7.	8.
9.	10.	11.	12.
13.	14.	15.	16.
17.	18.	19.	20.

Lesson #135

1. What number is 65% of 160?

2. What is the value of x? $-2x + 6 = -x$

3. $1.642 \div 0.02 = ?$

4. $93 + (-52) = ?$

5. Write the equation for the line through points $(-8, 2)$ and $(1, 3)$.

6. Find the value of x. $7x - 11 = 10$

7. $\sqrt{100} - \sqrt{9} = ?$

8. $k - 13 > -11$ Graph on a number line.

9. Solve for x. $\dfrac{x}{7} + 25 = 105$

10. $\dfrac{5}{6} \times \dfrac{18}{20} = ?$

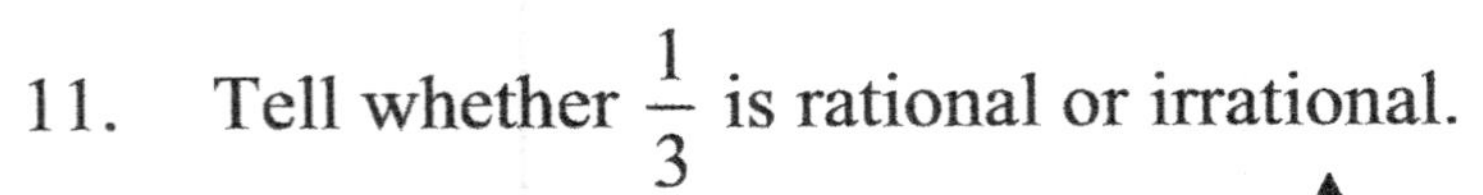

11. Tell whether $\dfrac{1}{3}$ is rational or irrational.

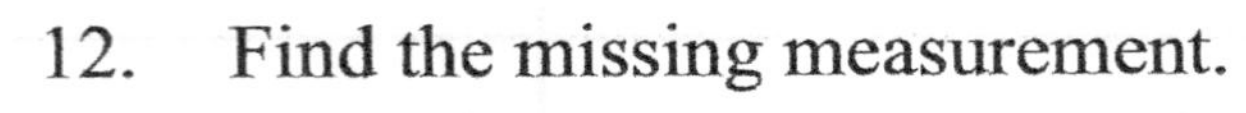

12. Find the missing measurement.

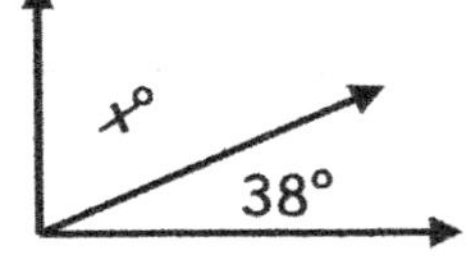

13. $-|-48| = ?$

14. $50 \div 5 + 4 \cdot 7 - 32 \div 8 = ?$

15. Combine like terms. $7a + 3ab + 9b - 10 - 2a + 2ab - 5$

16. Solve for x. $|x| + 4 = 15$

17. $16 - 8\dfrac{2}{5} = ?$

18. Write 1.24×10^{-3} in standard form.

19. Write an expression for *seventy-two plus a number.*

20. Write $\dfrac{4}{25}$ as a decimal and as a percent.

1.	2.	3.	4.
5.	6.	7.	8.
9.	10.	11.	12.
13.	14.	15.	16.
17.	18.	19.	20.

Lesson #136

1. Solve the inequality for h. $-22 \geq h - 12$

2. The scale on a map is 0.25 inch to 1.5 miles. The distance between Neston Falls and Langley on the map is 3 inches. What is the actual distance between the two towns?

3. Solve for b. $\frac{1}{12}b + 13 = 20$

4. What is the value of y? $y + 27 = -54$

5. Find the area of a triangle that is 20 cm wide at its base and 9 cm high.

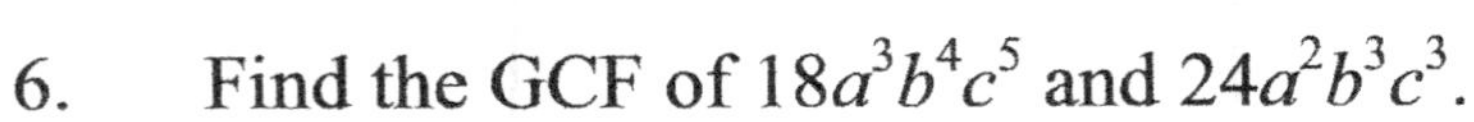

6. Find the GCF of $18a^3b^4c^5$ and $24a^2b^3c^3$.

7. Solve for x. $-5x = -105$

8. Write the slope-intercept form of a linear equation.

9. Water boils at _______°F.

10. If $a = 5$, $b = 9$ and $c = 3$, what is the value of $\frac{ab}{c} - 9$?

11. Solve for k. $4 + k > 3$

12. What is the P(1, 4) on 2 rolls of a die?

13. $-66 + (-39) = ?$

14. $\frac{3}{8} \times \frac{12}{15} = ?$

15. Write the equation for the line through points (3, 5) and (5, 3).

16. $3.624 \div 0.6 = ?$

17. How many feet are in 7 yards?

18. Write 82,000 in scientific notation.

19. 70% of 25 is what number?

20. $8\frac{1}{2} + 5\frac{3}{4} = ?$

1.	2.	3.	4.
5.	6.	7.	8.
9.	10.	11.	12.
13.	14.	15.	16.
17.	18.	19.	20.

Lesson #137

1. Solve the inequality for x. $-2x - 10 \geq 4$

2. $4\frac{1}{3} + 6\frac{2}{5} = ?$

3. Find the percent of change from 2.3 cm to 2.8 cm. Round to the nearest whole number.

4. What is the value of w? $w + 23 = 54$

5. Find the value of x. $\frac{1}{8}x = 14$

6. Write the equation for the line through points (25, 100) and (15, 120).

7. Simplify. $\frac{16x^4y^3z^3}{24xy^3z}$

8. How many feet are in 6 yards?

9. $-66 + (-48) = ?$

10. Solve the proportion for x. $\frac{8}{x} = \frac{40}{30}$

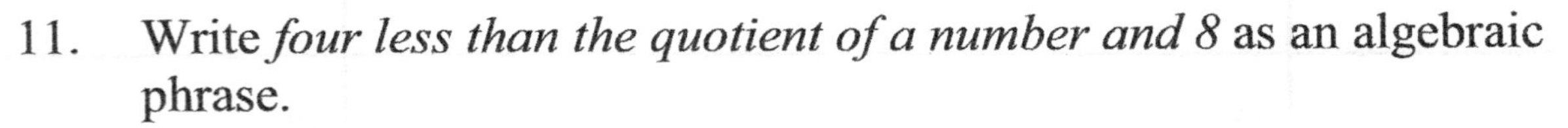

11. Write *four less than the quotient of a number and 8* as an algebraic phrase.

12. Solve for x. $4x + 17 = -7$

13. Find the GCF of $9a^3b^2c$ and $12a^2bc^2$.

14. Distribute and combine like terms. $4(2x + 2y - 7) - 3(3x - 4y + 2)$

15. Write the equation for the line through point (5, –8), with a slope of –3.

16. Solve for y. $5y + 3 = 23$

17. What is the value of x? $\frac{x}{12} + 9 = 20$

18. Find the perimeter of a decagon whose sides each measure 5 cm.

19. $-|-10| = ?$

20. $\sqrt{100} + \sqrt{81} - 2^3 = ?$

1.	2.	3.	4.
5.	6.	7.	8.
9.	10.	11.	12.
13.	14.	15.	16.
17.	18.	19.	20.

Lesson #138

1. $s + 6 \leq 5$ Graph the solution on a number line.

2. $-4(-2)(-3) = ?$

3. What is the value of x? $11x + 45 = 2x$

4. The slope of a horizontal line is ____________.

5. Write 0.76 as a percent and as a reduced fraction.

6. $-98 - (-36) = ?$

7. What is the P(1, 3, 2, 5) on 4 rolls of a die?

8. What is the slope of the line passing through points (–8, 0) and (1, 5)?

9. The sum of Rachel and Mary's ages is 20. Mary is 4 years younger than Rachel. What are their ages?

10. Find the value of x. $18x + 2x - 4 = 6$

11. A line has the equation, $y = \frac{7}{6}x + 3$. Find its slope and y-intercept.

12. Solve for c. $\frac{5c-8}{18} = \frac{2}{3}$

13. $36 + (-21) + (-26) = ?$

14. Solve for x. $x - 86 = -125$

15. $13\frac{1}{6} - 10\frac{5}{6} = ?$

16. $\begin{pmatrix} 6 & -8 \\ 4 & 6 \end{pmatrix} - \begin{pmatrix} -3 & 2 \\ 1 & -7 \end{pmatrix} = ?$

17. $12 + 3[4 + 8 - 3] = ?$

18. Write 7.33×10^{-2} in standard form.

19. $\frac{-565}{5} = ?$

20. Write the formula for finding the circumference of a circle.

1.	2.	3.	4.
5.	6.	7.	8.
9.	10.	11.	12.
13.	14.	15.	16.
17.	18.	19.	20.

Lesson #139

1. Solve for x. $5x - 3x + 7 = -13$
2. In science class, each student measured and recorded the height of a plant. The first week's measurements were $3\frac{1}{2}$ inches, $5\frac{2}{3}$ inches, $4\frac{1}{2}$ inches, $4\frac{3}{4}$ inches, and 5 inches. What is the median of this data?
3. Use algebraic symbols for *a number divided by six, decreased by nine.*
4. $135 + (-75) = ?$
5. Solve for s. $3(s - 4) \geq 6s + 12$
6. $30 \div 6 + 5 \cdot 4 - 12 \div 4 = ?$
7. Find the value of x. $\frac{x}{12} = -12$
8. Solve for x. $6x - 12 = 3x + 3$
9. Find the circumference of a circle whose diameter is 14 mm.
10. $2.3 - 1.75 = ?$
11. How many centuries are 800 years?
12. Write the formula for finding the volume of a cylinder.
13. Write $\frac{7}{25}$ as a decimal and as a percent.
14. Find $\frac{5}{6}$ of 42.
15. Write the slope-intercept form of a linear equation.
16. Solve the proportion for x. $\frac{18}{24} = \frac{12}{x}$
17. Evaluate $\frac{abc}{3}$ when $a = 2$, $b = 3$ and $c = 15$.
18. $24\frac{1}{3} - 12\frac{2}{3} = ?$
19. Find the missing measurement, x.

174°
x°

20. A team's win-loss ratio was 5 to 9. If the team won 15 games, how many did the team lose?

1.	2.	3.	4.
5.	6.	7.	8.
9.	10.	11.	12.
13.	14.	15.	16.
17.	18.	19.	20.

Lesson #140

1. Solve for n. $6(2n-5)=42$
2. $\frac{4}{5}\times\frac{20}{24}=?$
3. A triangle with three congruent sides is a(n) __________ triangle.
4. Find the median of 10, 45, 36, 89 and 77.
5. What is the value of x? $7x+12=13x$
6. $1.806\div 0.3=?$
7. Find the value of a. $-5a=70$
8. Write 93.4 in scientific notation.
9. $37+(-14)=?$
10. Simplify. $\frac{15x^2y}{21xy^2}$
11. 16% of 50 is what number?
12. How many gallons are 16 quarts?
13. Solve the inequality for c. $3(2c-4)\geq 48$
14. What is the probability of rolling *a number less than 5* on one roll of a die?
15. $\sqrt{169}=?$
16. Solve for x. $x+43=-96$
17. $\frac{8}{12}\div\frac{2}{6}=?$
18. -33 ◯ -49
19. Solve for x. $\frac{4}{5}x=-20$
20. Determine the slope of the line passing through points (0, – 4) and (2, –2).

1.	2.	3.	4.
5.	6.	7.	8.
9.	10.	11.	12.
13.	14.	15.	16.
17.	18.	19.	20.

Algebra I
Part A

Help Pages & "Who Knows"

Help Pages

Vocabulary

General

Absolute Value — the distance between a number, x, and zero on a number line; written as $|x|$.
Example: $|5| = 5$ reads "The absolute value of 5 is 5." $|-7| = 7$ reads "The absolute value of -7 is 7."

Expression — a mathematical phrase written in symbols. Example: $2x + 5$ is an expression.

Function — a rule that pairs each number in a given set (the domain) with just one number in another set (the range). Example: The function $y = x + 3$ pairs every number with another number that is larger by 3.

Greatest Common Factor (GCF) — the highest factor that 2 numbers have in common.
Example: The factors of 6 are 1, 2, **3** and 6. The factors of 9 are 1, **3** and 9. The GCF of 6 and 9 is 3.

Integers — the set of whole numbers, positive or negative, and zero.

Irrational number — a number that cannot be written as the ratio of two whole numbers. The decimal form of an irrational number is <u>neither</u> terminating nor repeating. Examples: $\sqrt{2}$ and π.

Least Common Multiple (LCM) — the smallest multiple that 2 numbers have in common.
Example: Multiples of 3 are 3, 6, 9, **12**, 15... Multiples of 4 are 4, 8, **12**, 16... The LCM of 3 and 4 is 12.

Matrix — a rectangular arrangement of numbers in rows and columns. Each number in a matrix is an element or entry. The plural of matrix is matrices. Example: $\begin{pmatrix} 2 & 3 \\ 0 & -1 \end{pmatrix}$ is a matrix with 4 elements.

Rational number — a number that can be written as the ratio of two whole numbers.
Example: 7 is rational; it can be written as $\frac{7}{1}$. 0.25 is rational; it can be written as $\frac{1}{4}$.

Slope — the ratio of the *rise* (vertical change) to the *run* (horizontal change) for a non-vertical line.

Square Root — a number that when multiplied by itself gives you another number. The symbol for square root is $\sqrt{x}$. Example: $\sqrt{49} = 7$ reads "The square root of 49 is 7."

Term — the components of an expression, usually being added to or subtracted from each other.
Example: The expression $2x + 5$ has two terms: $2x$ and 5. The expression $3n^2$ has only one term.

Geometry

Acute Angle — an angle measuring less than 90°.

Complementary Angles — two angles whose measures add up to 90°.

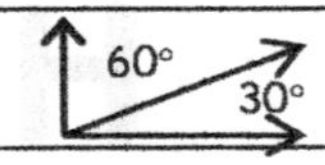

Congruent — figures with the same shape and the same size.

Obtuse Angle — an angle measuring more than 90°.

Right Angle — an angle measuring exactly 90°.

Similar — figures having the same shape but with different sizes.

Straight Angle — an angle measuring exactly 180°.

Supplementary Angles — two angles whose measures add up to 180°. 110° 70°

Surface Area — the sum of the areas of all of the faces of a solid figure.

Help Pages

Vocabulary (continued)

Geometry — Circles

Circumference — the distance around the outside of a circle.

Diameter — the widest distance across a circle. The diameter always passes through the center.

Radius — the distance from any point on the circle to the center. The radius is half of the diameter.

Geometry — Triangles

Equilateral — a triangle with all 3 sides having the same length.

Isosceles — a triangle with 2 sides having the same length.

Scalene — a triangle with none of the sides having the same length.

Geometry — Polygons

Number of Sides	Name	Number of Sides	Name
3	Triangle	7	Heptagon
4	Quadrilateral	8	Octagon
5	Pentagon	9	Nonagon
6	Hexagon	10	Decagon

Measurement — Relationships

Volume	Distance
3 teaspoons in a tablespoon	36 inches in a yard
2 cups in a pint	1760 yards in a mile
2 pints in a quart	5280 feet in a mile
4 quarts in a gallon	100 centimeters in a meter
Weight	1000 millimeters in a meter
16 ounces in a pound	**Temperature**
2000 pounds in a ton	0° Celsius – Freezing Point
Time	100°Celsius – Boiling Point
10 years in a decade	32°Fahrenheit – Freezing Point
100 years in a century	212°Fahrenheit – Boiling Point

Ratio and Proportion

Proportion — a statement that two ratios (or fractions) are equal. Example: $\frac{1}{2} = \frac{3}{6}$

Percent (%) — the ratio of any number to 100. Example: 14% means 14 out of 100 or $\frac{14}{100}$.

Help Pages

Solved Examples

Absolute Value

The **absolute value** of a number is its distance from zero on a number line. It is always positive.

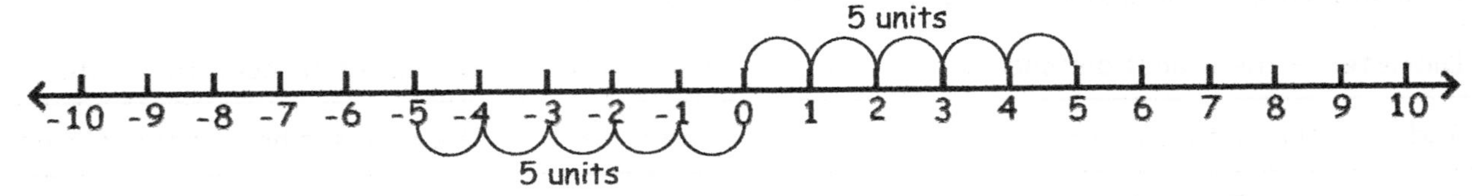

The absolute value of both -5 and +5 is 5, because both are 5 units away from zero. The symbol for the absolute value of -5 is |-5|. Examples: |-3| = 3; |8| = 8.

Equations

An equation consists of two expressions separated by an equal sign. You have worked with simple equations for a long time: $2+3=5$. More complicated equations involve variables which replace a number. To solve an equation like this, you must figure out which number the variable stands for. A simple example is when $2+x=5$, $x=3$. Here, the variable, x, stands for 3.

Sometimes an equation is not so simple. In these cases, there is a process for solving for the variable. No matter how complicated the equation, <u>the goal is to work with the equation until all the numbers are on one side and the variable is alone on the other side</u>. These equations will require only **one step** to solve. To check your answer, put the value of x back into the original equation.

Solving an **equation with a variable on one side**:

Example: Solve for x. $x + 13 = 27$

$$\begin{aligned} x+13 &= 27 \\ -13 &= -13 \\ \hline x \quad &= 14 \end{aligned}$$

1. Look at the side of the equation that has the variable on it. If there is a number added to or subtracted from the variable, it must be removed. In the first example, 13 is added to x.
2. To remove 13, add its opposite (-13) to both sides of the equation.
3. Add downward. x plus nothing is x. 13 plus -13 is zero. 27 plus -13 is 14.
4. Once the variable is alone on one side of the equation, the equation is solved. The bottom line tells the value of x. $x = 14$.

Example: Solve for a. $a - 22 = -53$

$$\begin{aligned} a-22 &= -53 \\ +22 &= +22 \\ \hline a \quad &= -31 \end{aligned}$$

Check: -31 -22 = -53 ✓ correct!

Help Pages

Solved Examples

Equations (continued)

In the next examples, a number is either multiplied or divided by the variable (not added or subtracted).

Example: Solve for x. $3x = 39$

$$3x = 39$$
$$\frac{3x}{3} = \frac{39}{3}$$
$$x = 13$$

Check: $3(13) = 39$
$39 = 39$ ✓ correct!

Example: Solve for n. $\frac{n}{6} = -15$

$$\frac{n}{\cancel{6}}(\cancel{6}) = -15(6)$$
$$n = -90$$

Check: $\frac{-90}{6} = -15$
$-15 = -15$ ✓ correct!

1. Look at the side of the equation that has the variable on it. If there is a number multiplied by or divided into the variable, it must be removed. In the first example, 3 is multiplied by x.
2. To remove 3, divide both sides by 3. (You divide because it is the opposite operation from the one in the equation (multiplication).
3. Follow the rules for multiplying or dividing integers. $3x$ divided by 3 is x. 39 divided by 3 is thirteen.
4. Once the variable is alone on one side of the equation, the equation is solved. The bottom line tells the value of x. $x = 13$.

The next set of examples also have a variable on only one side of the equation. These, however, are a bit more complicated, because they will require **two steps** in order to get the variable alone.

Example: Solve for x. $2x + 5 = 13$

$$2x + 5 = 13$$
$$-5 = -5$$
$$2x = 8$$
$$\frac{\cancel{2}x}{\cancel{2}} = \frac{8}{2}$$
$$x = 4$$

Check: $2(4) + 5 = 13$
$8 + 5 = 13$
$13 = 13$ ✓ correct!

Example: Solve for n. $3n - 7 = 32$

$$3n - 7 = 32$$
$$+7 = +7$$
$$3n = 39$$
$$\frac{\cancel{3}n}{\cancel{3}} = \frac{39}{3}$$
$$n = 13$$

Check: $3(13) - 7 = 32$
$39 - 7 = 32$
$32 = 32$ ✓ correct!

1. Look at the side of the equation that has the variable on it. There is a number (2) multiplied by the variable, and there is a number added to it (5). Both of these must be removed. Always begin with the addition/subtraction. To remove the 5 we must add its opposite(-5) to both sides.
2. To remove the 2, divide both sides by 2. (You divide because it is the opposite operation from the one in the equation (multiplication).
3. Follow the rules for multiplying or dividing integers. $2x$ divided by 2 is x. 8 divided by 2 is four.
4. Once the variable is alone on one side of the equation, the equation is solved. The bottom line tells the value of x. $x = 4$.

Help Pages

Solved Examples

Equations (continued)

These multi-step equations also have a variable on only one side. To get the variable alone, though, requires several steps.

Example: Solve for x. $3(2x+3)=21$

$$\frac{\not{3}(2x+3)}{\not{3}}=\frac{21}{3}$$
$$2x+3=7$$
$$-3=-3$$
$$2x=4$$
$$\frac{\not{2}x}{\not{2}}=\frac{4}{2}$$
$$x=2$$

Check: $3(2(2)+3)=21$
$3(4+3)=21$
$3(7)=21$
$21 = 21$ ✓ correct!

1. Look at the side of the equation that has the variable on it. First, the expression ($2x$+ 3) is multiplied by 3; then there is a number (3) added to $2x$, and there is a number (2) multiplied by x. All of these must be removed. To remove the 3 outside the parentheses, divide both sides by 3. (You divide because it is the opposite operation from the one in the equation (multiplication).
2. To remove the 3 inside the parentheses, add its opposite (-3) to both sides.
3. Remove the 2 by dividing both sides by 2.
4. Follow the rules for multiplying or dividing integers. $2x$ divided by 2 is x. 4 divided by 2 is two.
5. Once the variable is alone on one side of the equation, the equation is solved. The bottom line tells the value of x. x= 2.

When solving an **equation with a variable on both sides**, the goals are the same: to get the numbers on one side of the equation and to get the variable alone on the other side.

Example: Solve for x. $2x+4=6x-4$

$$2x+4=6x-4$$
$$-2x\quad=-2x$$
$$4=4x-4$$
$$+4=\quad+4$$
$$8=4x$$
$$\frac{8}{4}=\frac{\not{4}x}{\not{4}}$$
$$2=x$$

Check: $2(2) + 4 = 6(2) - 4$
$4 + 4 = 12 - 4$
$8 = 8$ ✓ correct!

1. Since there are variables on both sides, the first step is to remove the "variable term" from one of the sides by adding its opposite. To remove $2x$ from the left side, add $-2x$ to both sides.
2. There are also numbers added (or subtracted) to both sides. Next, remove the number added to the variable side by adding its opposite. To remove -4 from the right side, add +4 to both sides.
3. The variable still has a number multiplied by it. This number (4) must be removed by dividing both sides by 4.
4. The final line shows that the value of x is 2.

Help Pages

Solved Examples

Equations (continued)

Example: Solve for n. $5n - 3 = 8n + 9$

$$\begin{aligned} 5n - 3 &= 8n + 9 \\ -8n \quad &= -8n \\ \hline -3n - 3 &= 9 \\ +3 &= +3 \\ \hline -3n &= 12 \\ \frac{-3n}{-3} &= \frac{12}{-3} \\ n &= -4 \end{aligned}$$

Check: $5(-4) - 3 = 8(-4) + 9$

$-20 - 3 = -32 + 9$

$-23 = -23$ ✓ correct!

Exponents

An **exponent** is a small number to the upper right of another number (the base). Exponents are used to show that the base is a repeated factor.

Example: 2^4 is read "two to the fourth power."

base → 2^4 ← exponent

The base (2) is a factor multiple times.

The exponent (4) tells how many times the base is a factor.

$2^4 = 2 \times 2 \times 2 \times 2 = 16$

Example: 9^3 is read "nine to the third power" and means $9 \times 9 \times 9 = 729$

Expressions

An **expression** is a number, a variable, or any combination of these, along with operation signs $(+, -, \times, \div)$ and grouping symbols. An expression never includes an equal sign.

Five examples of expressions are 5, x, $(x + 5)$, $(3x + 5)$, and $(3x^2 + 5)$.

To **evaluate an expression** means to calculate its value using specific variable values.

Example: Evaluate $2x + 3y + 5$ when $x = 2$ and $y = 3$.

$2(2) + 3(3) + 5 = ?$

$4 + 9 + 5 = ?$

$13 + 5 = 18$

The expression has a value of 18.

1. To evaluate, put the values of x and y into the expression.
2. Use the rules for integers to calculate the value of the expression.

Example: Find the value of $\frac{xy}{3} + 2$ when $x = 6$ and $y = 4$.

$\frac{6(4)}{3} + 2 = ?$

$\frac{24}{3} + 2 = ?$

$8 + 2 = 10$

The expression has a value of 10.

Help Pages

Solved Examples

Expressions (continued)

Some expressions can be made more simple. There are a few processes for **simplifying an expression**. Deciding which process or processes to use depends on the expression itself. With practice, you will be able to recognize which of the following processes to use.

The **distributive property** is used when one term is multiplied by (or divided into) an expression that includes either addition or subtraction. $a(b+c)=ab+ac$ or $\frac{b+c}{a}=\frac{b}{a}+\frac{c}{a}$

Example: Simplify $3(2x+5)$.

$$3(2x+5)=$$
$$3(2x)+3(5)=$$
$$6x+15$$

1. Since the 3 is multiplied by the expression, $2x+5$, the 3 must be multiplied by both terms in the expression.
2. Multiply 3 by $2x$ and then multiply 3 by +5.
3. The result includes both of these: $6x+15$. Notice that simplifying an expression does not result in a single number answer, only a more simple expression.

Example: Simplify $2(7x-3y+4)$.

$$2(7x-3y+4)=$$
$$2(7x)+2(-3y)+2(+4)=$$
$$14x-6y+8$$

Expressions which contain like-terms can also be simplified. **Like-terms** are those that contain the same variable to the same power. $2x$ and $-4x$ are like-terms; $3n^2$ and $8n^2$ are like-terms; $5y$ and y are like-terms; 3 and 7 are like-terms.

An expression sometimes begins with like-terms. This process for **simplifying expressions** is called **combining like-terms**. When combining like-terms, first identify the like-terms. Then, simply add the like-terms to each other and write the results together to form a new expression.

Example: Simplify $2x+5y-9+5x-3y-2$.

The like-terms are $2x$ and $+5x$, $+5y$ and $-3y$, and -9 and -2.

$2x + +5x =$ **$+7x$**, $+5y + -3y =$ **$+2y$**, and $-9 + -2 =$ **-11**.

The result is **$7x + 2y - 11$**.

The next examples are a bit more complex. It is necessary to use the distributive property first, and then to combine like-terms.

Example: Simplify $2(3x+2y+2)+3(2x+3y+2)$

$$\begin{array}{r} 6x+4y+4 \\ +6x+9y+6 \\ \hline 12x+13y+10 \end{array}$$

Example: Simplify $4(3x-5y-4)-2(3x-3y+2)$

$$\begin{array}{r} +12x-20y-16 \\ -6x+6y-4 \\ \hline 6x-14y-20 \end{array}$$

1. First, apply the distributive property to each expression. Write the results on top of each other, lining up the like terms with each other. Pay attention to the signs of the terms.
2. Then, add each group of like-terms. Remember to follow the rules for integers.

Help Pages

Solved Examples

Expressions (continued)

Other expressions that can be simplified are written as fractions. **Simplifying** these expressions (**algebraic fractions**) is similar to simplifying numerical fractions. It involves cancelling out factors that are common to both the numerator and the denominator.

Simplify $\frac{12x^2yz^4}{16xy^3z^2}$.

$$\frac{\overset{3}{\cancel{12}}\ \overset{x}{\cancel{x^2}}\ \cancel{y}\ \overset{z^2}{\cancel{z^4}}}{\underset{4}{\cancel{16}}\ \cancel{x}\ \underset{y^2}{\cancel{y^3}}\ \cancel{z^2}}$$

$$\frac{\cancel{2}\cdot\cancel{2}\cdot 3\cdot\cancel{x}\cdot x\cdot\cancel{y}\cdot\cancel{z}\cdot\cancel{z}\cdot z\cdot z}{\cancel{2}\cdot\cancel{2}\cdot 2\cdot 2\cdot\cancel{x}\cdot\cancel{y}\cdot y\cdot y\cdot\cancel{z}\cdot\cancel{z}}$$

$$\frac{3xz^2}{4y^2}$$

1. Begin by looking at the numerals in both the numerator and denominator (12 and 16). What is the largest number that goes into both evenly? Cancel this factor (4) out of both.
2. Look at the x portion of both numerator and denominator. What is the largest number of x's that can go into both of them? Cancel this factor (x) out of both.
3. Do the same process with y and then z. Cancel out the largest number of each (y and z^2). Write the numbers that remain in the numerator or denominator for your answer.

Often a relationship is described using verbal (English) phrases. In order to work with the relationship, you must first **translate it into an algebraic expression or equation**. In most cases, word clues will be helpful. Some examples of verbal phrases and their corresponding algebraic expressions or equations are written below.

Verbal Phrase	Algebraic Expression
Ten more than a number	$x + 10$
The sum of a number and five	$x + 5$
A number increased by seven	$x + 7$
Six less than a number	$x - 6$
A number decreased by nine	$x - 9$
The difference between a number and four	$x - 4$
The difference between four and a number	$4 - x$
Five times a number	$5x$
Eight times a number, increased by one	$8x + 1$
The product of a number and six is twelve.	$6x = 12$
The quotient of a number and 10	$\frac{x}{10}$
The quotient of a number and two, decreased by five	$\frac{x}{2} - 5$

In most problems, the word "is" tells you to put in an equal sign. When working with fractions and percents, the word "of" generally means multiply. Look at the example below.

One half <u>of</u> a number <u>is</u> fifteen.

You can think of it as "one half <u>times</u> a number <u>equals</u> fifteen."

When written as an algebraic equation, it is $\frac{1}{2}x = 15$.

Help Pages

Solved Examples

Expressions (continued)

At times you need to find the **Greatest Common Factor (GCF) of an algebraic expression.**

Example: Find the GCF of $12x^2yz^3$ and $18xy^3z^2$.

1. First, find the GCF of the numbers (12 and 18). The largest number that is a factor of both is **6**.
2. Now look at the x's. Of the x-terms, which contains fewest x's. Comparing x^2 and x, x contains the fewest.
3. Now look at the y's and then the z's. Again, of the y-terms, y contains the fewest. Of the z-terms, z^2 contains the fewest.
4. The GCF contains all of these: $\mathbf{6xyz^2}$.

$12x^2yz^3$ and $18xy^3z^2$

The GCF of 12 and 18 is **6**.

Of x^2 and x, the smallest is x.

Of y and y^3, the smallest is y.

Of z^3 and z^2, the smallest is z^2.

The GCF is: $\mathbf{6xyz^2}$.

At other times you need to know the **Least Common Multiple (LCM) of an algebraic expression.**

Example: Find the LCM of $10a^3b^2c^2$ and $15ab^4c$.

1. First, find the LCM of the numbers (10 and 15). The lowest number that both go into evenly is **30**.
2. Now look at the a-terms. Which has the largest number of a's. Comparing a^3 and a, a^3 has the most.
3. Now look at the b's and then the c's. Again, of the b-terms, b^4 contains the most. Of the c-terms, c^2 contains the most.
4. The LCM contains all of these: $\mathbf{30a^3b^4c^2}$.

$10a^3b^2c^2$ and $15ab^4c$

The LCM of 10 and 15 is **30**.

Of a^3 and a, the largest is a^3.

Of b^2 and b^4, the largest is b^4.

Of c^2 and c, the largest is c^2.

The LCM is: $\mathbf{30a^3b^4c^2}$.

Functions

A **function** is a rule that pairs each number in a given set (the domain) with just one number in another set (the range). A function performs one or more operations on an input-number which results in an output-number. The set of all input-numbers is called the **domain** of the function. The set of all output-numbers is called the **range** of the function. Often, a function table is used to help organize your thinking.

Example: For the function, $y = 3x$, find the missing numbers in the function table.

x	y
2	?
-1	?
?	15

The function is $y = 3x$. This function multiplies every x-value by 3.

When we input $x = 2$, we get $y = 3(2)$ or $y = 6$. →

When we use $x = -1$, we get $y = 3(-1)$ or $y = -3$. →

When we use $y = 15$, we get $15 = 3x$, so $\frac{15}{3} = x$ or $5 = x$. →

x	y
2	6
-1	-3
5	15

The set of all inputs is the domain. For this function table, the domain is {2, -1, 5}

The set of all outputs is the range. For this function table, the range is {6, -3, 15}.

Help Pages

Solved Examples

Geometry

To find the **area of a triangle**, first recognize that any triangle is exactly half of a parallelogram.

The whole figure is a parallelogram.

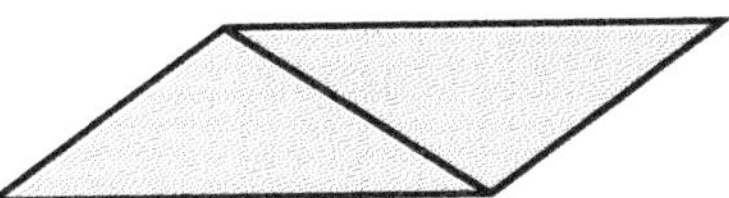

Half of the whole figure is a triangle.

So, the triangle's area is equal to half of the product of the base and the height.

Area of triangle = $\frac{1}{2}$(base × height) or $A = \frac{1}{2}bh$

Examples: Find the area of the triangles below.

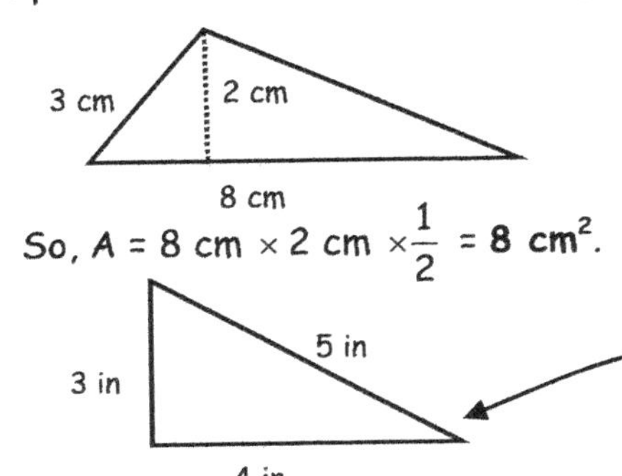

So, $A = 8\text{ cm} \times 2\text{ cm} \times \frac{1}{2} = \mathbf{8\ cm^2}$.

1. Find the length of the base. (8 cm)
2. Find the height. (It is 2cm. The height is always straight up and down – never slanted.)
3. Multiply them together and divide by 2 to find the area. (8 cm^2)

The base of this triangle is 4 inches long. Its height is 3 inches. (Remember the height is always straight up and down!)

So, $A = 4\text{ in} \times 3\text{ in} \times \frac{1}{2} = \mathbf{6\ in^2}$.

Finding the **area of a parallelogram** is similar to finding the area of any other quadrilateral. The area of the figure is equal to the length of its base multiplied by the height of the figure.

Area of parallelogram = base × height or $A = b \times h$

Example: Find the area of the parallelogram below.

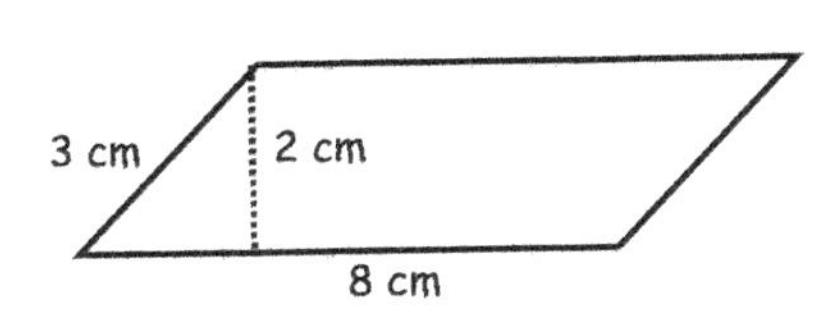

1. Find the length of the base. (8 cm)
2. Find the height. (It is 2cm. The height is always straight up and down – never slanted.)
3. Multiply to find the area. (16 cm^2)

So, $A = 8\text{ cm} \times 2\text{ cm} = \mathbf{16\ cm^2}$.

Geometry (continued)

Finding the **area of a trapezoid** is a little different than the other quadrilaterals that we have seen. Trapezoids have 2 bases of unequal length. To find the area, first find the average of the lengths of the 2 bases. Then, multiply that average by the height.

$$\text{Area of trapezoid} = \frac{\text{base}_1 + \text{base}_2}{2} \times \text{height} \quad \text{or} \quad A = \left(\frac{b_1 + b_2}{2}\right)h$$

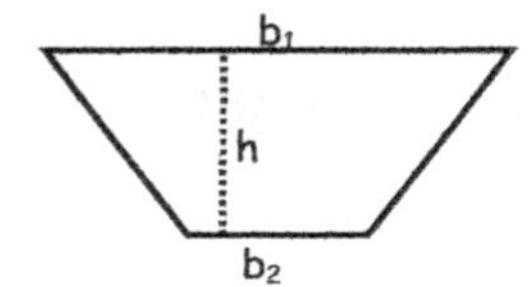

The bases are labeled b_1 and b_2.

The height, h, is the distance between the bases.

Examples: Find the area of the trapezoid below.

1. Add the lengths of the two bases. (22 cm)
2. Divide the sum by 2. (11 cm)
3. Multiply that result by the height to find the area. (110 cm^2)

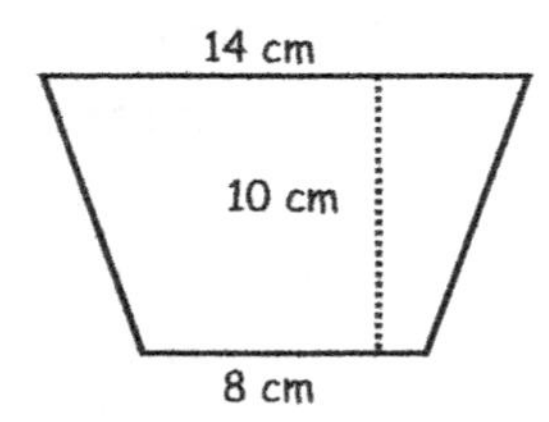

$$\frac{14\,\text{cm} + 8\,\text{cm}}{2} = \frac{22\,\text{cm}}{2} = 11\,\text{cm}$$

11 cm × 10 cm = **110 cm^2** = Area

To **find the measure of an angle**, a protractor is used.

The symbol for angle is $\angle$. On the diagram, $\angle AOE$ has a measure less than 90°, so it is acute.

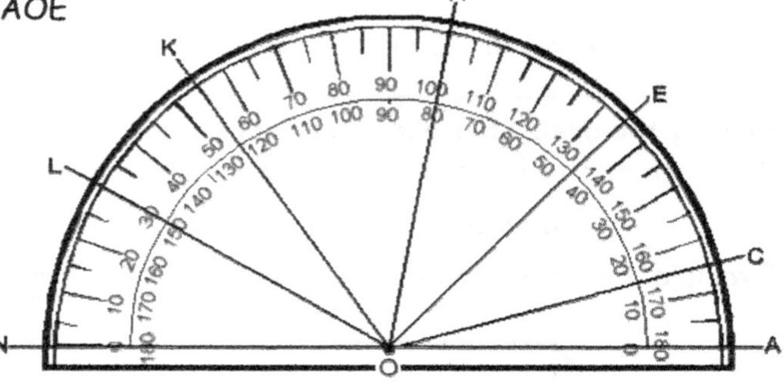

With the center of the protractor on the vertex of the angle (where the 2 rays meet), place one ray ($\overrightarrow{OA}$) on one of the "0" lines. Look at the number that the other ray ($\overrightarrow{OE}$) passes through. Since the angle is acute, use the lower set of numbers. Since $\overrightarrow{OE}$ is halfway between the 40 and the 50, the measure of $\angle AOE$ is 45°. (If it were an obtuse angle, the higher set of numbers would be used.)

Look at $\angle NOH$. It is an obtuse angle, so the higher set of numbers will be used. Notice that $\overrightarrow{ON}$ is on the "0" line. $\overrightarrow{OH}$ passes through the 100 mark. So the measure of $\angle NOH$ is 100°.

Help Pages

Solved Examples

Geometry (continued)

The **circumference of a circle** is the distance around the outside of the circle. Before you can find the circumference of a circle you must know either its radius or its diameter. Also, you must know the value of the constant, *pi* (π). $\pi = 3.14$ (rounded to the nearest hundredth).

Once you have this information, the circumference can be found by multiplying the diameter by *pi*.

Circumference = $\pi \times$ diameter or $C = \pi d$

Examples: Find the circumference of the circles below.

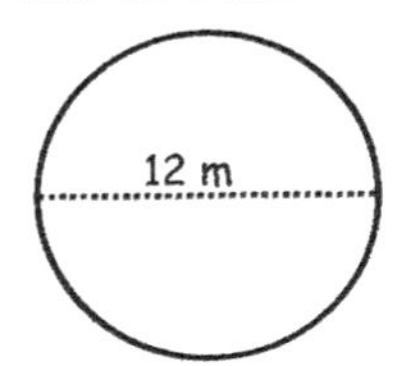

1. Find the length of the diameter. (12 m)
2. Multiply the diameter by π. (12m $\times$ 3.14)
3. The product is the circumference. (37.68 m)

So, C = 12 m $\times$ 3.14 = **37.68 m.**

Sometimes the radius of a circle is given instead of the diameter. Remember, the radius of any circle is exactly half of the diameter. If a circle has a radius of 3 feet, its diameter is 6 feet.

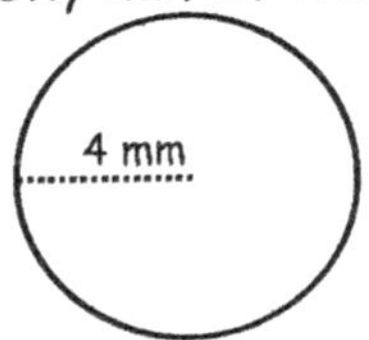

Since the radius is 4 mm, the diameter must be 8 mm.

Multiply the diameter by π. (8 mm $\times$ 3.14)

The product is the circumference. (25.12 mm)

So, C = 8 mm $\times$ 3.14 = **25.12 mm.**

When finding the **area of a circle**, the length of the radius is squared (multiplied by itself), and then that answer is multiplied by the constant, *pi* (π). $\pi = 3.14$ (rounded to the nearest hundredth).

Area = $\pi \times$ radius $\times$ radius or $A = \pi r^2$

Examples: Find the area of the circles below.

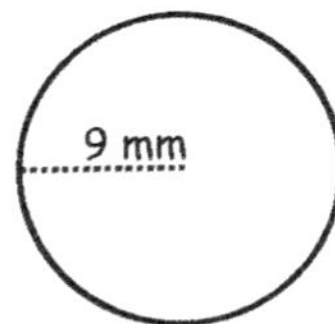

1. Find the length of the radius. (9 mm)
2. Multiply the radius by itself. (9 mm x 9 mm)
3. Multiply the product by *pi*. (81 mm^2 x 3.14)
4. The result is the area. (254.34 mm^2)

So, A = 9 mm x 9 mm x 3.14 = **254.34 mm^2.**

Sometimes the diameter of a circle is given instead of the radius. Remember, the diameter of any circle is exactly twice the radius. If a circle has a diameter of 6 feet, its radius is 3 feet.

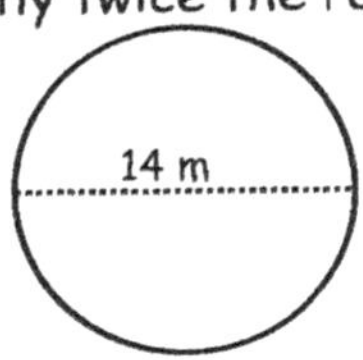

Since the diameter is 14 m, the radius must be 7 m.

Square the radius. (7 m x 7 m)

Multiply that result by π. (49 m^2 $\times$ 3.14)

The product is the area. (153.86 m^2)

So, A = (7 m)^2 $\times$ 3.14 = **153.86 m^2.**

Help Pages

Solved Examples

Geometry (continued)

To find the **surface area** of a solid figure, it is necessary to first count the total number of faces. Then, find the area of each of the faces; finally, add the areas of each face. That sum is the surface area of the figure.

Here, the focus will be on finding the **surface area of a rectangular prism**. A rectangular prism has 6 faces. Actually, the opposite faces are identical, so this figure has 3 pairs of faces. Also, a prism has only 3 dimensions: Length, Width, and Height.

This prism has identical left and right sides (A & B), identical top and bottom (C & D), and identical front and back (unlabeled).

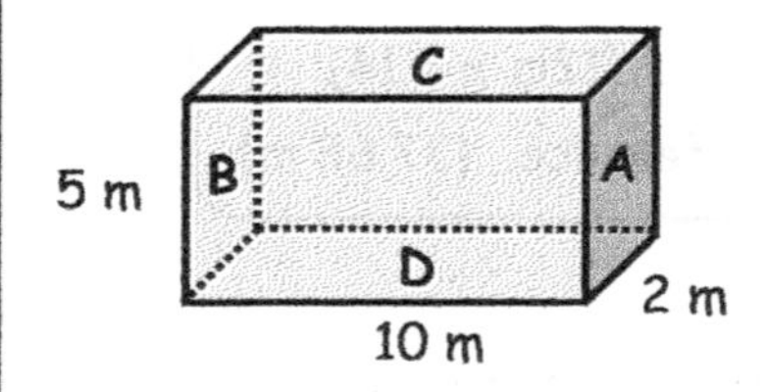

1. Find the area of the front: L x W. ($10 \text{ m} \times 5 \text{ m} = 50 \text{ m}^2$)
   Since the back is identical, its area is the same.
2. Find the area of the top (C): L x H. ($10 \text{ m} \times 2 \text{ m} = 20 \text{ m}^2$)
   Since the bottom (D) is identical, its area is the same.
3. Find the area of side A: W x H. ($2 \text{ m} \times 5 \text{ m} = 10 \text{ m}^2$) Since side B is identical, its area is the same.
4. Add up the areas of all 6 faces.
   ($10 \text{ m}^2 + 10 \text{ m}^2 + 20 \text{ m}^2 + 20 \text{ m}^2 + 50 \text{ m}^2 + 50 \text{ m}^2 = \mathbf{160 \text{ m}^2}$)

Surface Area of a Rectangular Prism = 2(length x width) + 2(length x height) + 2(width x height)

or $SA = 2LW + 2LH + 2WH$

To find the **volume** of a solid figure, it is necessary to determine the area one face and multiply that by the height of the figure. Volume of a solid is measured in cubic units (cm^3, in^3, ft^3, etc.).

Here the focus will be on finding the **volume of a cylinder**. As shown below, a cylinder has two identical circular faces.

Example: Find the volume of the cylinder below.

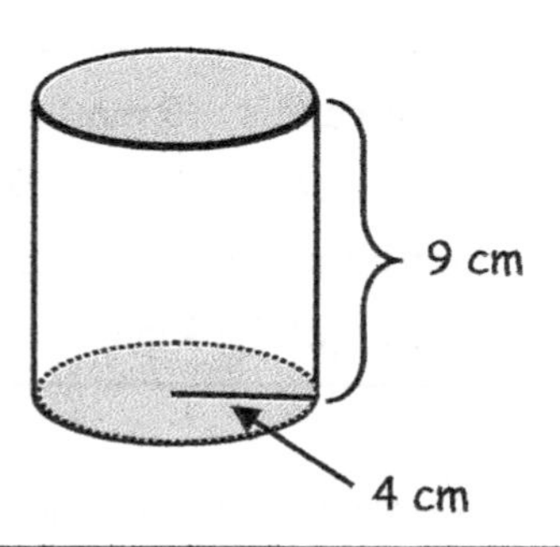

1. To find the area of one of the circular faces, multiply the constant, π (3.14), by the square of the radius (4 cm). Area = $3.14 \times (4 \text{ cm})^2 = 50.24 \text{ cm}^2$
2. The height of this cylinder is 9 cm. Multiply the height by the area calculated in Step 1.
   Volume = $50.24 \text{ cm}^2 \times 9 \text{ cm} = 452.16 \text{ cm}^3$

Help Pages

Solved Examples

Graphing

A **coordinate plane** is formed by the intersection of a horizontal number line, called the ***x*-axis**, and a vertical number line, called the ***y*-axis**. The axes meet at the point (0, 0), called the **origin**, and divide the coordinate plane into four **quadrants**.

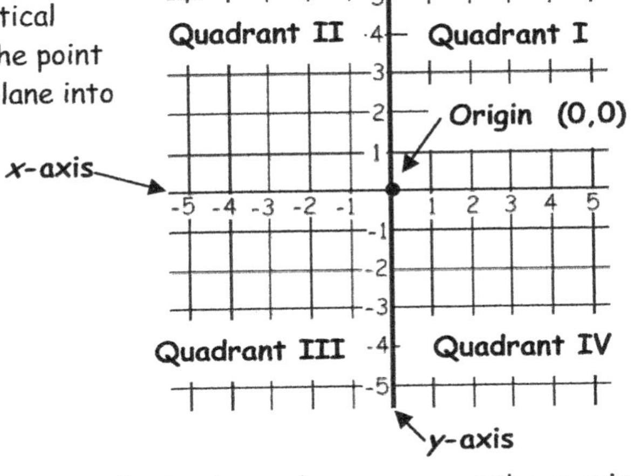

Points are represented by **ordered pairs** of numbers, (x, y). The first number in an ordered pair is the x-coordinate; the second number is the y-coordinate. In the point **(-4, 1)**, -4 is the x-coordinate and 1 is the y-coordinate.

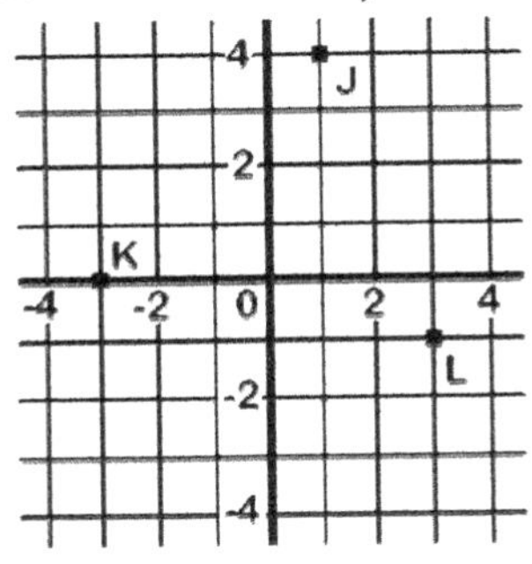

When graphing on a coordinate plane, always move on the x-axis first (right or left), and then move on the y-axis (up or down).

- The coordinates of point J are (1, 4).
- The coordinates of point K are (-3, 0).
- The coordinates of point L are (3, -1).

On a coordinate plane, any 2 points can be connected to form a line. The line, however, is made up of many points – in fact, every place on the line is another point. One of the properties of a line is its slope (or steepness). The **slope** of a non-vertical line is the ratio of its vertical change (rise) to its horizontal change (run) between any two points on the line. The slope of a line is represented by the letter m. Another property of a line is the ***y*-intercept**. This is the point where the line intersects the y-axis. A line has only one y-intercept, which is represented by the letter b.

$$\text{Slope of a line} = \frac{\text{change in } y}{\text{change in } x} = \frac{\text{rise}}{\text{run}}$$

The rise-over-run method can be used to find the slope if you're looking at the graph.

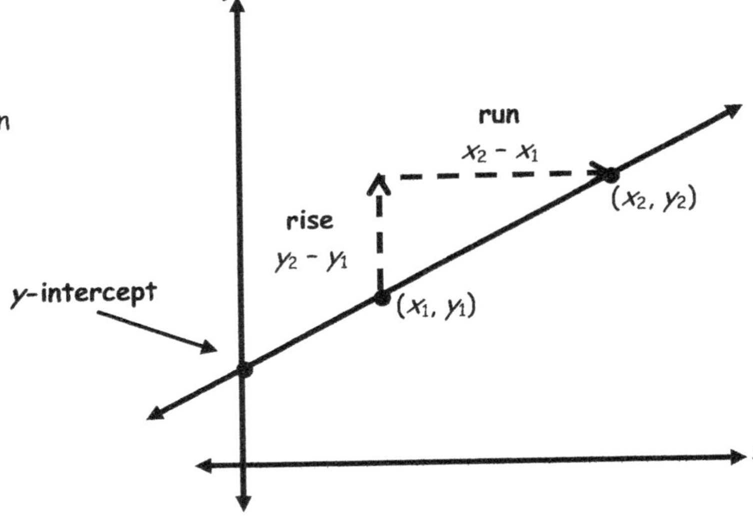

Help Pages

Solved Examples

Graphing (continued)

Another way to find the slope of a line is to use the formula. The formula for slope is $m = \frac{y_2 - y_1}{x_2 - x_1}$, where the two points are (x_1, y_1) and (x_2, y_2).

Example: What is the slope of $\overline{AD}$?

Point A with coordinates (3, 4) and Point D with coordinates (1, 2) are both on this line.

For Point A, x_2 is 3 and y_2 is 4.

For Point D, x_1 is 1 and y_1 is 2.

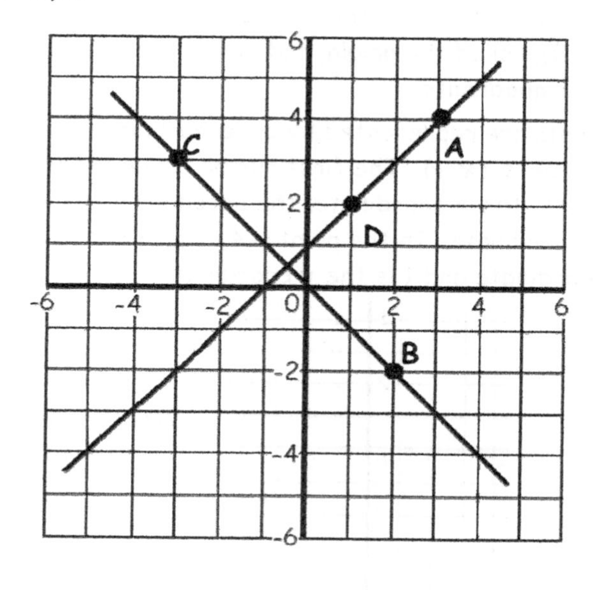

slope = $m = \frac{y_2 - y_1}{x_2 - x_1} = \frac{4-2}{3-1} = \frac{2}{2} = 1$

The slope of $\overline{AD}$ is 1.

Use the formula to find the slope of $\overline{CB}$.

Point C is (-3, 3) and Point B is (2, -2).

slope = $m = \frac{y_2 - y_1}{x_2 - x_1} = \frac{-2-3}{2-(-3)} = \frac{-5}{5} = -1$

The slope of $\overline{CB}$ is -1.

Every line has an equation which describes it, called a linear equation. We will focus on one particular form of linear equation - **slope-intercept form**. To write the slope-intercept equation of a line, you must know the slope and the y-intercept.

A linear equation in slope-intercept form is always in the form $y = mx + b$, where m is the slope, b is the y-intercept, and (x, y) is any point on the line.

Example: A line has the equation $y = 2x + 5$. What is the slope? What is the y-intercept?

$y = 2x + 5$

$y = mx + b$

The slope, m, is 2. The y-intercept, b, is 5.

Example: A line has a slope of 6 and a y-intercept of -3. Write the equation for the line.

The slope is 6, so m = 6. The y-intercept is -3, so b = -3.

Put those values into the slope-intercept form: $y = 6x - 3$

Example: Write the equation of a line that passes through points (3, 2) and (6, 4).

Only 2 things are needed to write the equation of a line: slope and y-intercept.

First, find the slope. $m = \frac{y_2 - y_1}{x_2 - x_1} = \frac{4-2}{6-3} = \frac{2}{3}$

Then, find the y-intercept. Choose either point. Let's use (6, 4). The x-value of this point is 6 and the y-value is 4. Put these values along with the slope into the equation and solve for b.

$y = mx + b$ $\quad 4 = \frac{2}{3}(6) + b$ $\quad 4 = \frac{12}{3} + b$ $\quad 4 = 4 + b$ $\quad 0 = b$

So the slope = $\frac{2}{3}$ and the y-intercept = 0. The equation of the line is $y = \frac{2}{3}x + 0$.

Help Pages

Solved Examples

Inequalities

An **inequality** is a statement that one quantity is different than another (usually larger or smaller). The symbols showing inequality are $<, >, \leq$, and $\geq$. (less than, greater than, less than or equal to, and greater than or equal to). An inequality is formed by placing one of the inequality symbols between two expressions. The solution of an inequality is the set of numbers that can be substituted for the variable to make the statement true.

A simple inequality is $x \leq 4$. The solution set, {..., 2, 3, 4}, includes all numbers that are either less than four or equal to four.

Some inequalities are solved using only addition or subtraction. The approach to solving them is similar to that used when solving equations. The goal is to get the variable alone on one side of the inequality and the numbers on the other side.

Examples: Solve $x - 4 < 8$.

$$\begin{array}{r} x - 4 < 8 \\ +4 \quad +4 \\ \hline x \quad < 12 \end{array}$$

1. To get the variable alone, add the opposite of the number that is with it to both sides.
2. Simplify both sides of the inequality.
3. Graph the solution on a number line. For $<$ and $>$, use an open circle; for $\leq$ and $\geq$, use a closed circle.

Solve $y + 3 \geq 10$.

$$\begin{array}{r} y + 3 \geq 10 \\ -3 \quad -3 \\ \hline y \quad \geq 7 \end{array}$$

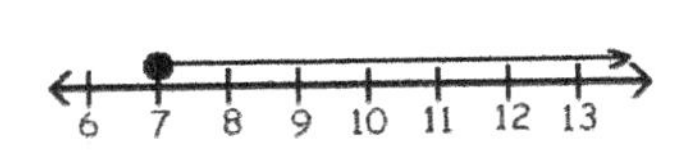

Some inequalities are solved using only multiplication or division. The approach to solving them is also similar to that used when solving equations. Here, too, the goal is to get the variable alone on one side of the inequality and the numbers on the other side.

The one difference that you must remember is this: If, when solving a problem you multiply or divide by a negative number, you must flip the inequality symbol.

Examples: Solve $8n < 56$.

$$\frac{8n}{8} < \frac{56}{8}$$

$$n < 7$$

1. Check to see if the variable is being multiplied or divided by a number.
2. Use the same number, but do the opposite operation on both sides.
3. Simplify both sides of the inequality.
4. Graph the solution on a number line. For $<$ and $>$, use an open circle; for $\leq$ and $\geq$, use a closed circle.

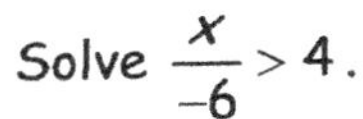

Solve $\frac{x}{-6} > 4$.

$$\frac{x}{-6} > 4$$

$$(-6)\frac{x}{-6} < 4(-6)$$

$$x < -24$$

Notice that during the 2nd step, when multiplying by -6, the sign "flipped" from greater than to less than.

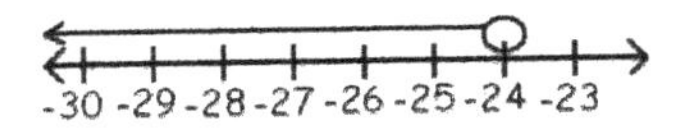

REMEMBER: When multiplying or dividing an inequality by a negative number, the inequality symbol must be flipped!

Help Pages

Solved Examples

Inequalities (continued)

Some inequalities must be solved using both addition/subtraction and multiplication/division. In these problems, the addition/subtraction is always done first.

Example: $2x-6\le 6$

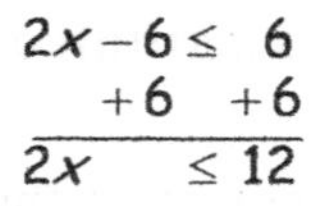

$$\begin{aligned} 2x-6 &\le 6 \\ +6 &\quad +6 \\ 2x &\le 12 \\ \frac{2x}{2} &\le \frac{12}{2} \\ x &\le 6 \end{aligned}$$

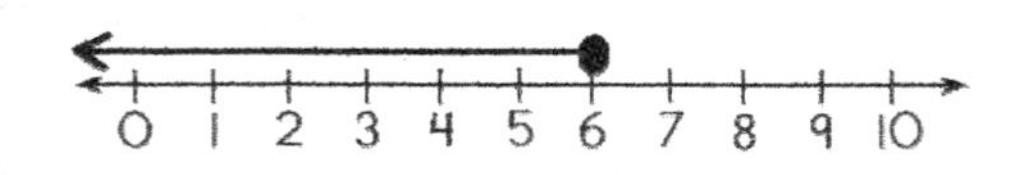

A compound inequality is a statement comparing one quantity (in the middle) with two other quantities (on either side).

$-2<y<1$ This can be read "y is greater than -2, but less than 1."

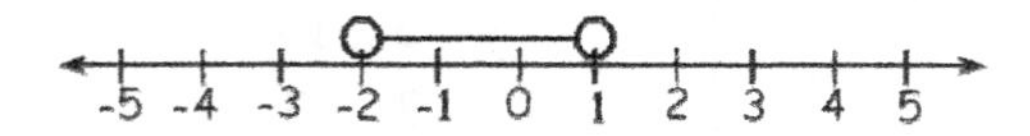

Integers

Integers include the counting numbers, their opposites (negative numbers) and zero.

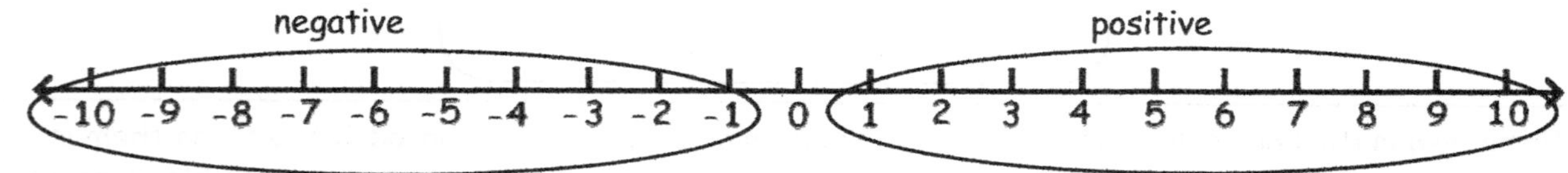

The negative numbers are to the left of zero. The positive numbers are to the right of zero.

When **ordering integers**, they are being arranged either from least to greatest or from greatest to least. The further a number is to the right, the greater its value. For example, 9 is further to the right than 2, so 9 is greater than 2.

In the same way, -1 is further to the right than -7, so -1 is greater than -7.

Examples: Order these integers from **least to greatest**: -10, 9, -25, 36, 0

Remember, the smallest number will be the one farthest to the left on the number line, -25, then -10, then 0. Next will be 9, and finally 36.

Answer: -25, -10, 0, 9, 36

Put these integers in order from **greatest to least**: -94, -6, -24, -70, -14

Now the greatest value (farthest to the right) will come first and the smallest value (farthest to the left) will come last.

Answer: -6, -14, -24, -70, -94

Help Pages

Solved Examples

Integers

The rules for performing operations $(+,-,\times,\div)$ on integers are very important and must be memorized.

The Addition Rules for Integers:

1. When the signs are the same, add the numbers and keep the sign. When the signs are different, subtract the numbers and use the sign of the larger number.

$$\begin{array}{r}+33\\++19\\\hline+52\end{array}\qquad\begin{array}{r}-23\\+-19\\\hline-39\end{array}\qquad\begin{array}{r}+35\\+-21\\\hline+14\end{array}\qquad\begin{array}{r}-55\\++27\\\hline-28\end{array}$$

The Subtraction Rule for Integers:

Change the sign of the second number and add (follow the Addition Rule for Integers above).

$$\begin{array}{r}+56\\--26\\\hline\end{array}\ \xrightarrow{\text{apply rule}}\ \begin{array}{r}+56\\++26\\\hline+82\end{array}\qquad\begin{array}{r}+48\\-+23\\\hline\end{array}\ \xrightarrow{\text{apply rule}}\ \begin{array}{r}+48\\+-23\\\hline+25\end{array}$$

Notice that every subtraction problem becomes an addition problem, using this rule!

The Multiplication and Division Rule for Integers:

1. When the signs are the same, the answer is positive (+).

$+7\times+3=+21$ $\quad$ $-7\times-3=+21$

$+18\div+6=+3$ $\quad$ $-18\div-6=+3$

2. When the signs are different, the answer is negative (-).

$+7\times-3=-21$ $\quad$ $-7\times+3=-21$

$-18\div+6=-3$ $\quad$ $+18\div-6=-3$

+	×	+	=	+
–	×	–	=	+
+	×	–	=	–
–	×	+	=	–
+	÷	+	=	+
–	÷	–	=	+
+	÷	–	=	–
–	÷	+	=	–

Matrix, Matrices

A **matrix** is a rectangular arrangement of numbers in rows and columns. Each number in a matrix is an element or entry. The plural of matrix is **matrices**.

$$\begin{pmatrix}0&4&-1\\-3&2&5\end{pmatrix}$$

The matrix to the right has 2 rows and 3 columns. It has 6 elements.

In order to be added or subtracted, matrices must have the same number of rows and columns. If they don't have the same dimensions, they cannot be added or subtracted.

When **adding matrices**, simply add corresponding elements.

Example: $$\begin{pmatrix}0&4&-1\\-3&2&5\end{pmatrix}+\begin{pmatrix}2&1&3\\-2&-6&4\end{pmatrix}=\begin{pmatrix}(0+2)&(4+1)&(-1+3)\\(-3+(-2))&(2+(-6))&(5+4)\end{pmatrix}=\begin{pmatrix}2&5&2\\-5&-4&9\end{pmatrix}$$

When subtracting matrices, remember the subtraction rule for integers. A simple way to subtract matrices is to change the signs of every element of the second matrix. Then change the operation to addition and follow the rule for addition of integers (as shown in the previous example).

Example: $$\begin{pmatrix}-10&2\\3&-7\end{pmatrix}-\begin{pmatrix}5&-3\\6&-1\end{pmatrix}=$$

First change all signs, then add.

$$\begin{pmatrix}-10&2\\3&-7\end{pmatrix}+\begin{pmatrix}-5&+3\\-6&+1\end{pmatrix}=\begin{pmatrix}(-10+(-5))&(2+3)\\(3+(-6))&(-7+1)\end{pmatrix}=\begin{pmatrix}-15&+5\\-3&-6\end{pmatrix}$$

Help Pages

Solved Examples

Proportion

A **proportion** is a statement that two ratios are equal to each other. There are two ways to solve a proportion when a number is missing.

1. One way to solve a proportion is already familiar to you. You can use the equivalent fraction method.

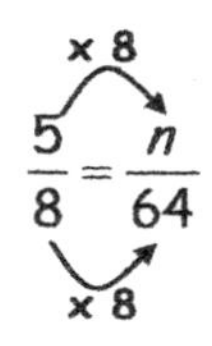

$$\frac{5}{8} = \frac{n}{64} \quad (\times 8)$$

$n = 40$.

So, $\frac{5}{8} = \frac{40}{64}$.

2. Another way to solve a proportion is by using cross-products.

To use Cross-Products:

1. Multiply downward on each diagonal.
2. Make the product of each diagonal equal to each other.
3. Solve for the missing variable.

$$\frac{14}{20} = \frac{21}{n}$$

$$20 \times 21 = 14 \times n$$

$$420 = 14n$$

$$\frac{420}{14} = \frac{14n}{14}$$

$$30 = n$$

So, $\frac{14}{20} = \frac{21}{30}$.

Percent

When changing from a fraction to a percent, a decimal to a percent, or from a percent to either a fraction or a decimal, it is very helpful to use an FDP chart (Fraction, Decimal, Percent).

To change a **fraction to a percent and/or decimal**, first find an equivalent fraction with 100 in the denominator. Once you have found that equivalent fraction, it can easily be written as a decimal. To change that decimal to a percent, move the decimal point 2 places to the right and add a % sign.

Example: Change $\frac{2}{5}$ to a percent and then to a decimal.

1. Find an equivalent fraction with 100 in the denominator.
2. From the equivalent fraction above, the decimal can easily be found. Say the name of the fraction: "forty hundredths." Write this as a decimal: 0.40.
3. To change 0.40 to a percent, move the decimal two places to the right. Add a % sign: 40%.

$$\frac{2}{5} = \frac{?}{100} \quad (\times 20) \qquad ? = 40$$

$$\frac{2}{5} = \frac{40}{100} = 0.40$$

$$0.40 = 40\%$$

When changing from a **percent to a decimal or a fraction**, the process is similar to the one used above. Write the percent as a fraction with a denominator of 100; reduce this fraction. Return to the percent, move the decimal point 2 places to the left. This is the decimal.

Example: Write 45% as a fraction and then as a decimal.

1. Begin with the percent. (45%) Write a fraction where the denominator is 100 and the numerator is the "percent." $\frac{45}{100}$
2. This fraction must be reduced. The reduced fraction is $\frac{9}{20}$.
3. Go back to the percent. Move the decimal point two places to the left to change it to a decimal.

$$45\% = \frac{45}{100}$$

$$\frac{45(\div 5)}{100(\div 5)} = \frac{9}{20}$$

$$45\% = .45$$

Decimal point goes here.

Help Pages

Solved Examples

Percent (continued)

When changing from a **decimal to a percent or a fraction**, again, the process is similar to the one used above. Begin with the decimal. Move the decimal point 2 places to the right and add a % sign. Return to the decimal. Write it as a fraction and reduce.

Example: Write 0.12 as a percent and then as a fraction.

1. Begin with the decimal. (0.12) Move the decimal point two places to the right to change it to a percent.
2. Go back to the decimal and write it as a fraction. Reduce this fraction.

$0.12 = 12\%$

0.12 = twelve hundredths

$$= \frac{12}{100} = \frac{12(\div 4)}{100(\div 4)} = \frac{3}{25}$$

Percent of change shows how much a quantity has increased or decreased from its original amount. When the new amount is greater than the original amount, the percent of change is called the **percent of increase**. When the new amount is less than the original amount, the percent of change is called the **percent of decrease**. Both of these are found in the same way. The difference between the new amount and the original amount is divided by the original amount. The result is multiplied by 100 to get the percent of change.

Formula: $$\%\text{ of change} = \frac{\text{amount of increase or decrease}}{\text{original amount}} \times 100$$

Example: A sapling measured 23 inches tall when it was planted. Two years later the sapling was 36 inches tall. What was the percent of increase? Round your answer to a whole number.

$$\left(\frac{36-23}{23}\right)\times 100 =$$

$$\left(\frac{13}{23}\right)\times 100 = 0.565$$ The sapling's height increased by 57% over the 2 years.

$$0.565 \times 100 = 57\%$$

Probability

The **probability of two or more independent events** occurring together can be determined by multiplying the individual probabilities together. The product is called the **compound probability**.

$$P(A \text{ and } B) = P(A) \times P(B)$$

Example: What is the probability of rolling a 6 and then a 2 on two rolls of a die [P(6 and 2)]?

A) First, since there are 6 numbers on a die and only one of them is a 6, the probability of getting a 6 is $\frac{1}{6}$.

B) Since there are 6 numbers on a die and only one of them is a 2, the probability of getting a 2 is $\frac{1}{6}$.

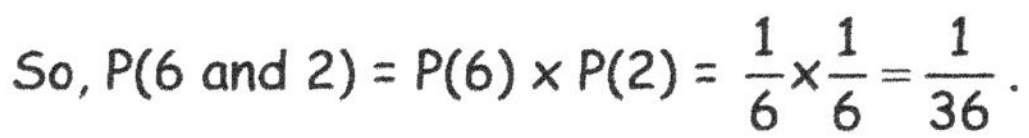

So, $P(6 \text{ and } 2) = P(6) \times P(2) = \frac{1}{6} \times \frac{1}{6} = \frac{1}{36}$.

There is a 1 to 36 chance of getting a 6 and then a 2 on two rolls of a die.

Help Pages

Solved Examples

Probability (continued)

Example: What is the probability of getting a 4 and then a number greater than 2 on two spins of this spinner [P(4 and greater than 2)]?

A) First, since there are 4 numbers on the spinner and only one of them is a 4, the probability of getting a 4 is $\frac{1}{4}$.

B) Since there are 4 numbers on the spinner and two of them are greater than 2, the probability of getting a 2 is $\frac{2}{4}$.

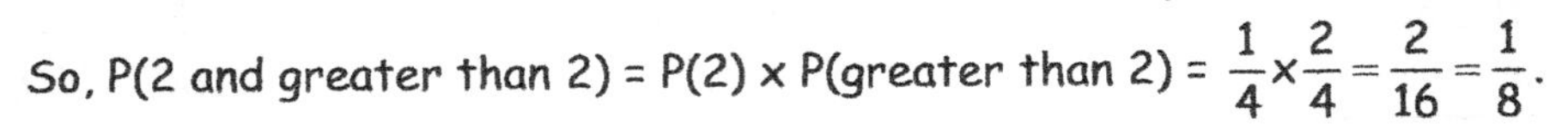

So, P(2 and greater than 2) = P(2) x P(greater than 2) = $\frac{1}{4} \times \frac{2}{4} = \frac{2}{16} = \frac{1}{8}$.

There is a 1 to 8 chance of getting a 4 and then a number more than 2 on two .

Example: On three flips of a coin, what is the probability of getting heads, tails, heads [P(H,T,H)]?

A) First, since there are only 2 sides on a coin and only one of them is heads, the probability of getting heads is $\frac{1}{2}$.

B) Again, there are only 2 sides on a coin and only one of them is tails. The probability of getting tails is also $\frac{1}{2}$.

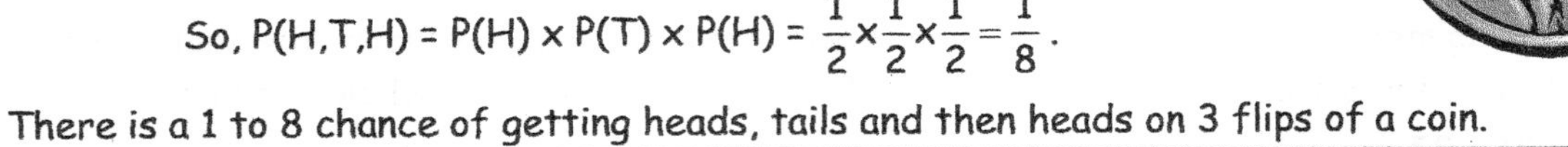

So, P(H,T,H) = P(H) x P(T) x P(H) = $\frac{1}{2} \times \frac{1}{2} \times \frac{1}{2} = \frac{1}{8}$.

There is a 1 to 8 chance of getting heads, tails and then heads on 3 flips of a coin.

Scientific Notation

Scientific notation is a shorthand method for representing numbers that are either very large or very small – numbers that have many zeroes and are tedious to write out.

For example, 5,000,000,000 and 0.000000023 have so many zeroes that it is not convenient to write them this way. Scientific notation removes the "placeholder" zeroes and represents them as powers of 10.

Numbers in scientific notation always have the form $c \times 10^n$ where $1 \le c < 10$ and n is an integer.

Examples: $5{,}000{,}000{,}000 = 5 \times 10^9$ $\quad$ $0.000000023 = 2.3 \times 10^{-8}$

5,000,000,000

5 000,000,000.

5×10^9

The decimal point was moved 9 places to the left, so the exponent is +9.

1. First locate the decimal point. Remember, if the decimal point isn't shown, it is after the last digit on the right.
2. Move the decimal point (either left or right) until the number is at least 1 and less than 10.
3. Count the number of places you moved the decimal point. This is the exponent.
4. If you moved the decimal to the right, the exponent will be negative; if you moved it to the left, the exponent will be positive.
5. Write the number times 10 to the power of the exponent that you found.

0.000000023

0.00000002 3

2.3×10^{-8}

The decimal point was moved 8 places to the right, so the exponent is -8.

Who Knows???

Degrees in a right angle? .... (90°)

A straight angle? ....(180°)

Angle greater than 90°? .... (obtuse)

Less than 90°? ....(acute)

Sides in a quadrilateral? ....(4)

Sides in an octagon? ....(8)

Sides in a hexagon? ....(6)

Sides in a pentagon? ....(5)

Sides in a heptagon? ....(7)

Sides in a nonagon? ....(9)

Sides in a decagon? ....(10)

Inches in a yard? .... (36)

Yards in a mile? ....(1,760)

Feet in a mile? .... (5,280)

Centimeters in a meter? .... (100)

Teaspoons in a tablespoon? ....(3)

Ounces in a pound? ....(16)

Pounds in a ton? ....(2,000)

Cups in a pint? ....(2)

Pints in a quart? ....(2)

Quarts in a gallon? ....(4)

Millimeters in a meter? ....(1,000)

Years in a century? .... (100)

Years in a decade? ....(10)

Celsius freezing? ....(0°C)

Celsius boiling? .... (100°C)

Fahrenheit freezing? ....(32°F)

Fahrenheit boiling? .... (212°F)

Number with only 2 factors? .... (prime)

Perimeter? ....(add the sides)

Area? ....(length x width)

Volume?.............. (length x width x height)

Area of parallelogram? ..... (base x height)

Area of triangle?............($\frac{1}{2}$ base x height)

Area of trapezoid ($\frac{base_1 + base_2}{2}$ x height)

Surface Area of a rectangular prism......... ....$SA = 2(LW) + 2(WH) + 2(LH)$

Volume of a cylinder .... ($\pi r^2 h$)

Area of a circle?.................................(πr^2)

Circumference of a circle? ....($d\pi$)

Triangle with no sides equal? ...... (scalene)

Triangle with 3 sides equal? .. (equilateral)

Triangle with 2 sides equal? ..... (isosceles)

Distance across the middle of a circle? ....(diameter)

Half of the diameter?.................... (radius)

Figures with the same size and shape? .... (congruent)

Figures with same shape, different sizes? .... (similar)

Number occurring most often?....... (mode)

Middle number? ....(median)

Answer in addition? ....(sum)

Answer in division?....................... (quotient)

Answer in multiplication? .... (product)

Answer in subtraction?...........(difference)